Human Reproduction:

Principles, Practices, Policies

Human Reproduction: Principles Practices Policies

Christine Overall

Toronto
OXFORD UNIVERSITY PRESS
1993

Oxford University Press, 70 Wynford Drive, Don Mills, Ontario M3C 1J9

Toronto Oxford New York
Delhi Bombay Calcutta Madras Karachi Kuala Lumpur
Singapore Hong Kong Tokyo Nairobi Dar es Salaam
Cape Town Melbourne Auckland Madrid

and associated companies in
Berlin Ibadan

This book is printed on permanent (acid-free) paper ∞

Canadian Cataloguing in Publication Data

Overall, Christine, 1949–
  Human reproduction : principles, practices, policies

Includes bibliographical references and index.
ISBN 0-19-540961-2

1. Human reproduction – Moral and ethical aspects.
I. Title.

QP251.084 1993   176   C93-094562-X

Cover Photograph: PHOTOTAKE (2) / FIRST LIGHT

Design: Marie Bartholomew

1 2 3 4 – 96 95 94 93

Printed in Canada

Contents

For Ted

Acknowledgements

An earlier version of 'The Co-optation of Feminist Values in Defence of Reproductive Engineering: A Case Study', was first printed under the title 'The Misuse of Feminist Values in the Defence of Reproductive Engineering: A Case Study', in *Resources for Feminist Research/Documentation sur la recherche féministe* 18, 3 (September 1989): 67-71, and is reprinted here with permission of the Coordinating Editor, Philinda Masters.

An earlier version of 'The Case Against Legalization of Contract Motherhood' was first printed in *Debating Canada's Future: Views From the Left*, edited by Simon Rosenblum and Peter Findlay, published by James Lorimer, 1991 (pp. 210-25), and is reprinted here with permission of James Lorimer.

An earlier version of 'Access to *In Vitro* Fertilization: Costs, Care, and Consent' was first printed in *Dialogue* 30, 2 (Summer 1991): 383-97, and is reprinted here with permission of the editor of *Dialogue*, Steven Davis.

An earlier version of 'Reproductive Engineering and Genealogy', entitled 'New Ways of Producing Babies: What is the Effect on Genealogy?' was first printed in *Family Tree Magazine* 8, 4 (February 1992): 28-9 and 8, 5 (March 1992): 28-30, and is reprinted here with permission of the editor, J.M. Armstrong.

Introduction

Recent research in the area of reproductive ethics—the philo-sophical study of moral and social-policy issues pertaining to procreative technologies and practices—has given ample confirm-ation, if any more was needed, of the political meanings and implications of personal experiences. My own work on these issues has been in part the outcome of experiences that turned out to have a more general political significance.

During the 1970s I was first a student and then a teacher of philosophy, an area dominated then as now by male academics. I was also a committed radical feminist, convinced that women's biol-ogy—particularly our sexual and reproductive capacities—is central to an understanding of women's position in patriarchal society.

That conviction was reinforced during my pregnancies. Para-doxically, the experience of carrying a foetus inside my body profoundly deepened my understanding of the abortion issue in two respects. First, the sensations of the active, unpredictable foetus with its turns and kicks, its hiccups and its rare rest peri-ods, convinced me that this entity within my body was definite-ly independent and alive, no mere appendage, and not a part of my body in the way that an arm or an eye or a heart is. At the same time, the experience of being 'with child' was sufficiently engrossing, disturbing, even overpowering at times, to persuade me that no woman should ever have to go through this experience—an experience that philosopher Caroline Whitbeck has suggested is akin to literally being possessed or taken over by another being[1]—against her will.

During the birth of my first child I was subjected to many technological interventions in labour and delivery, including intravenous stimulation of contractions, epidural anaesthesia, and internal foetal monitoring, primarily because my baby was considered to be 'overdue'. During my second delivery, the nurses ignored me after my arrival at the hospital because, despite my claims, they did not believe that I was in labour. Indeed, when my husband went to tell a nurse that I thought I was in transition (the brief period just before the pushing stage), the nurse replied, 'Tell her she just feels that way because she's breathing too hard.'

What, I wondered, is the significance of imposing major medical interventions on a healthy labouring woman? And why would what a woman says and feels about the birth of her own child be discounted and ignored?

During my first years of motherhood I learned, through firsthand experience, that although pregnancy and childbirth are significant and transformative events in women's lives, they are seldom seriously acknowledged or discussed as part of the public world of adult activities. I learned, too, that ours is a society that devalues procreation and mothering, regards mothers as ignorant or incompetent, and often treats babies and children as nuisances to whom no special concessions or facilities are owed. Despite lip service to the value and charm of children, they are treated as alien, as 'other', as non-persons without much moral status or importance in a culture where ageism is considered normal and being little is regarded as a handicap.

As a result of these experiences I thought there might be some interesting connections between issues related to abortion, about which philosophers had written a great deal since the revival of applied ethics in the early 1960s, and issues related to birthing, about which they had written virtually nothing. But in the late seventies the absence of philosophical writing on birth and reproduction made it difficult to believe that reproduction constituted a genuine subject for academic attention.

Then, in the early 1980s, one of my former students was discovered to have clear-cell adenocarcinoma, a cancerous condition of her reproductive system. Harriet Simand had been a brilliant student, graduating from college at the top of her class and entering McGill University on a full scholarship. The cause of her cancer was attributed to the fact that her mother had taken diethylstilbestrol (DES), a synthetic hormone, during pregnancy twenty years earlier. In late 1981 Harriet Simand underwent extensive gynaecological surgery, which excised the cancer but left her permanently sterile. She subsequently used the hard lessons she had learned from this experience to found DES Action/Canada, a support and advocacy group for DES mothers, daughters, and sons.

For me, the political significance of what had formerly appeared to be 'just' personal experiences connected with reproduction was now undeniable. The treatment of pregnant women and mothers, their foetuses and infants, had to be understood within the context of more general social responses to women and children. Human reproduction is not merely a set of individual biological processes, nor a private family event, but is determined by and through complex social interactions reflecting cultural values about the nature and value of baby-making. Procreation is not just a matter of human nature; it is a culturally constructed aspect of human life, and different cultures make different choices about how reproduction will be enacted and regulated. In Western culture, the medical establishment and the pharmaceutical industry do not necessarily promote the best interests of women and children; medical reactions to and treatments of pregnancy, birth, and childrearing incorporate certain value assumptions about the malleability and adequacy of women's bodies; and serious ethical and social policy questions are raised by new reproductive technologies (NRTs) and the social uses to which they are put. Thus issues in health-care ethics must be understood and assessed with reference to the wider political and social context in which they occur.

Since the late 1970s, when I first became interested in issues in reproductive ethics and policy, scientific practices and technologies

affecting procreation have burgeoned. These practices and technologies include the establishment of chains of clinics devoted to
the preselection of the sex of offspring; the development of a new
form of very early prenatal diagnosis by means of the examination
of the fertilized egg; the extensive therapeutic use of foetal tissue
to treat certain human diseases; the growth of embryo[2] experimentation; the increasing participation by women and heterosexual
couples in formal and informal contract motherhood arrangements; the expansion of proposed medical criteria for the use of
in vitro fertilization (IVF); long-term cryopreservation of fertilized
eggs, with subsequent disputes about entitlements to use them;
and the use of selective termination to reduce multiple pregnancies generated through fertility drugs and *in vitro* fertilization.

The scientific ideology used to justify the drive to expand and
perfect reproductive engineering incorporates a number of ideas
and assumptions. It is animated, first, by a focus on the enhance-
ment of fertility and by a view of procreation as central to, even
definitive of, women and womanhood. It assumes that virtually all
women should be mothers, and that wanting a baby is an inherent drive present in every 'normal' woman. At the same time,
women's bodies are seen as frail, awkward, imperfectly functioning mechanisms, whose operation can be improved and whose
product—infants—can be perfected through reproductive engineering. The result is a scientific and medical campaign to do everything possible to enable women to have babies. And not just
any babies: reproductive engineering holds out the promise of
'designer babies', whose sex can be chosen and whose physical
health can be assured through procedures involving the filtering
of sperm, the manipulation of embryos, and the examination of
foetuses. Thus, under this new medico-scientific regime, the child
becomes a commodity, a consumer durable with a variety of
optional characteristics, which is available for purchase to those
with sufficient money, time, and stamina.

◆ ◆ ◆
Reproductive Autonomy

Given the powerful influence of culture on procreative practices,
a central theme of this book is the exploration of the nature of

and conditions for women's reproductive autonomy and procreative freedom, and the ways in which many social, cultural, moral, and psychological norms and restrictions serve to undermine them.

At almost any age, women are profoundly affected by the availability or absence of reproductive choices. Just consider the seventeen-year-old women who is trying to get access to a safe, efficient means of contraception—or 'taking chances' and hoping each month that she hasn't got caught. Or the twenty-seven-year-old woman, pregnant for the second time and facing a repeat Caesarean delivery. Or the thirty-seven-year-old who has postponed childbearing for years and now finds that she can't conceive. Or the forty-seven-year-old, overwhelmed by guilt feelings because the drug she was prescribed during pregnancy years ago has produced serious reproductive problems in her twenty-year-old daughter.

Women in Canada today appear to enjoy a substantially higher degree of reproductive autonomy than ever before. Improvements in contraceptive technology, in the availability and safety of abortion services, and in medical care at birth seem to mean that women can choose not only how many children to have but when to have them, and, more radically still, whether to have children at all. Now, with the availability of techniques of prenatal diagnosis, sex preselection, and *in vitro* fertilization and cryopreservation, it seems that women can even choose what kinds of children to have.

Nonetheless, this apparent expansion of reproductive choice is not an unmixed blessing. Choices about our procreative future have consequences, some of which can be foreseen while others cannot, and not all of them are desirable. The intrauterine device (IUD), the contraceptive pill, and the cervical cap all give women more options for birth planning, but may also cause a variety of health problems for many of their users. The availability of electronic foetal monitoring during labour provides a kind of technological eye on the foetus, but its use may result in an increase in the number of Caesarean deliveries. The woman who chooses to have her children in her late thirties rather than in her twenties may benefit from the opportunity to build job skills and career experience, but runs a somewhat greater risk of diminished fertility and possible congenital abnormalities in her offspring. And the use of high-tech reproductive interventions such as IVF promises the woman with blocked or absent fallopian tubes the chance of having a baby, but is very seldom successful.

Moreover, the range of our reproductive choices still falls far short of the ideal, and the choices we now have should not be allowed to obscure those we have not attained. How freely can a woman choose motherhood when there is still no very adequate provision for maternity leave with pay from her job, or for day-care when she returns to work? How freely can a lesbian or a single heterosexual woman postpone motherhood if she may be considered ineligible at a later date for donor insemination (DI)[3] just because she is single? How freely can any woman in Canada plan her reproductive life when her access to abortion is depen-dent upon her geographical location? The papers in this volume investigate both the values that underlie current limitations on women's reproductive choices and the conditions that are required to further genuine procreative autonomy for women and for men.

♦ ♦ ♦
Moral Perspectives on Reproductive Issues

It is a truism that ethical discourse, moral decision-making, and social-policy formation have largely failed to keep pace with the growth of reproductive engineering. Nevertheless, the increasing medical interventions in the processes of procreation have been attended by an evolution in the moral frameworks brought to bear on the issues. Generally speaking, three very broad moral frameworks are evident in discussions of reproductive ethics; they may be called the conservative, liberal, and feminist perspectives.

The conservative approach to reproductive issues, exemplified by the Vatican encyclical 'Instruction on Respect for Human Life in Its Origin and on the Dignity of Procreation: Replies to Cer-tain Questions of the Day',[4] usually develops from religious premises, and emphasizes 'traditional' moral values. The conser-vative approach is founded upon essentialist views—derived from beliefs about the natural or divine order—of women, men, and human sexuality: views that assume there are innate characteristics of women, men, and sexual interactions, independent of the social context in which they exist, that make them what they are. Accenting the importance and maintenance of the nuclear family, and of heterosexuality and marriage as norms of adult sexual relationships, the conservative is characteristically suspicious of reproductive technologies, except in so far as their use is taken to promote the traditional family and conventional sexual and reproductive behaviour. From this point of view, then, some reproductive technologies and practices must be seen as being 'contrary to nature'. Most procreative innovations, from donor insemination to contract motherhood, are rejected as violating true sexual and family morality. Conservatives usually favour limitations on access even to those reproductive technologies they approve, based on characteristics such as marital status and sexual orientation, along with the outright banning or criminalization of other procedures and practices such as abortion. For the conservative, human personhood and full moral status begin at conception; hence the conservative perspective insists upon valuing and preserving embryos and foetuses, and is strongly opposed to practices such as embryo experimentation and foetal-tissue transplantation.

What I call the liberal approach to reproductive issues[5] places its emphasis on the moral dignity of the individual, making little differentiation between the situation of women and the situation of men in procreative practices. The liberal rejects the conservative view that human personhood and full moral status begin at conception, placing the landmark either later in gestational development or at birth. Hence the liberal is not opposed to abortion, embryo experimentation, or foetal-tissue transplantation. The liberal advocates the advancement of reproductive freedom, along with the promotion of individual rights, especially the right to privacy and the right to reproduce. Liberals usually have a positive attitude towards reproductive technology, arguing that for the most part it enhances reproductive freedom, and they evince confidence in scientific innovation and interventions in reproduction. In keeping with their emphasis on liberty, liberals wish to minimize limitations on the availability of and access to reproductive technologies; they believe that legal interventions are needed only to regulate the provision of procreative technologies and services. In addition, they rely largely on the publicity and partial regulation provided by the free market to enhance reproductive choice by permitting the buying and selling of gametes and embryos and the hiring of the services of contract mothers.

There are many forms and varieties of feminism, and feminists do not speak with one voice on the contentious issues raised by reproductive technologies and practices. Nonetheless, feminist approaches to reproductive issues usually have in common the following elements: a focus on women's lives, beliefs, values, experiences, and needs; a concern for women's reproductive health; and a regard for the well-being of children. Feminists see reproduction within the whole lifespan of women; reproductive issues have both a moral and a practical continuity with other social issues such as education for sexuality, prevention of AIDS, care of children, and mothering. Feminists have emphasized the need for a range of input on reproductive policy and ethics from women of all races, ethnicities, religions, and classes, and from persons with disabilities, persons who are infertile, and lesbians and gays.

Critical both of pronatalism—the cultural endorsement and encouragement of childbearing—and of views of women's nature that require a commitment to biological determinism, feminists argue that women should not be defined in terms of their capacity to bear children, or be limited to childrearing roles. Feminists are committed to a political analysis of reproductive behaviour, seeing reproductive issues not as isolated problems but as challenges that develop within the broader context of sexism, racism, homophobia and lesbophobia, and ableism. Like liberals, feminists have criticized discrimination in access to reproductive technologies and supported reproductive choice, yet they also remain highly critical of the actual and potential harmful effects of existing technologies and practices, and they raise questions about the conditions under which genuine reproductive choice can be fostered. They emphasize the importance not just of informed *consent* to medical procedures, but of active, informed *choice*, in which individuals are able to select from real alternatives.

Feminists have emphasized the importance of the study and prevention of infertility, as an alternative to heroic interventions in women's bodies after infertility has developed. While many texts cite high and growing rates of infertility to justify the increasing prevalence of reproductive engineering, it is important to remember that the evidence for such claims is ambiguous, and there may be good reasons for scepticism about the hand-wringing that often accompanies reports on infertility. While certain social and cultural changes (the growing incidence of sexually transmitted diseases; the use of dangerous contraceptives and pregnancy drugs such as DES; and environmental threats to reproductive health) suggest that the causes of infertility may be increasing, there is no adequate data base from early in the century to establish comparison rates of infertility. Moreover, infertility should not be conflated with sterility. Infertility is usually defined as the inability to conceive during one (or sometimes two) years of unprotected heterosexual intercourse. Yet there is no mention in this definition of how often intercourse takes place, or whether it occurs when the woman is ovulating. And the one- or two-year limit seems somewhat arbitrary: some women who by this definition are infertile will conceive after more than one or two years of trying.

Apart from these crucial areas of agreement, however, there are also significant disagreements among feminist theorists and activists. Feminists dispute the extent to which it is possible to talk about women as a general category, and the degree to which women's needs, especially *vis-à-vis* procreation, are socially constructed. There are also significant divergences among feminists about the legitimate extent of reproductive choice. Should it include, for example, access to *in vitro* fertilization for all women? Should it include the services of a contract mother? Should it include the freedom to determine the disposition of the foetus after abortion?

Feminists also disagree about the usefulness and moral integrity of appeals to the concept of moral rights in the construction of arguments in reproductive ethics. And even those who are willing to use the language of rights do not always agree about what rights to claim.

Perhaps most important, feminists disagree about the seriousness of various reproductive issues, the potential usefulness of reproductive technologies, the extent to which the proliferation of reproductive technologies is dangerous for women, and the degree to which (if at all) science, medicine, and the state can be used to advance women's goals and interests. Some, such as liberal feminists Lori Andrews and Carmel Shalev,[6] believe that most reproductive technologies should be easily available to women, whereas radical feminists such as Gena Corea and Barbara Katz Rothman are sceptical about the value, safety, and effectiveness of NRTs, and support limitations on such practices as contract motherhood.[7] Some feminists have called for a moratorium on any further developments in reproductive technologies, so that serious efforts can be made to assess the threats they may pose to the health of women and children.

◆ ◆ ◆

An Outline of the Chapters

Like those who adopt a liberal perspective on reproductive issues, I attribute central importance to the moral concepts of reproductive autonomy and reproductive rights. Yet at the same time I adopt a feminist interpretation of these ideals, since I am profoundly suspicious, where liberals are often complacent, about the proliferation of reproductive technologies, the harms of reproductive engineering, and the manipulation of reproductive needs, interests, and goals.

This book is a collection of papers written between 1988 and 1992, after the publication of my first book, *Ethics and Human Reproduction: A Feminist Analysis.*[8] Some of these papers evolved from my thinking about new reproductive issues and policies arising in the late 1980s and early 1990s; others developed in response to perceived gaps and problems in the views I formulated in the earlier work. Emphasizing the Canadian social context of these issues, this book attempts both to evaluate controversial reproductive practices and policies individually, and to illuminate some of their interconnections.

Chapter 1, 'Reflections on Reproductive Rights in Canada', is an overview of the ways in which the moral concepts of the right to reproduce and the right not to reproduce are currently interpreted within the Canadian political arena. In my view, it is at least premature and perhaps politically dangerous for progressive theorists and activists to abandon the moral language of rights at a time when similar language may be used to limit women's reproductive freedom. Moreover, the thoughtful and aware use of the language of rights is not necessarily inconsistent with the reconceptualization of our social, moral, and ontological institutions and relationships, including the relationship between the pregnant woman and the foetus. As a result, many of the papers in this volume make use of the fundamental distinctions, laid out in the first chapter, between the right to reproduce and the right not to reproduce, and the two senses, weak and strong, in which each of these may be interpreted. I suggest that there are two forms of the right to reproduce: a weak (or liberty) right not to be interfered with in procreative behaviour, and a strong (or welfare) right to assistance in reproduction. Recognition of the first form of the right to reproduce requires protection from forced sterilization, forced abortion, forced contraception, and racist marriage laws. Recognition of the second right, however, would seem to require provision of extensive reproductive services and technologies, such as *in vitro* fertilization, embryo sex preselection, embryo cryopreservation, and contract motherhood.

The justification for recognition of a right can be assessed by considering the possible consequences, good or bad, if the right is not recognized. While recognition of the first form of the right to reproduce is essential to the protection of people's fundamental reproductive autonomy, recognition of the second is less easily justified, since it would seem to allow violation of the rights of others. While some people may 'need' the gametes and gestational services of other people in order to reproduce, there can be no *right* to receive these gametes and services.

There are also two forms of the right not to reproduce. The right not to reproduce in the weak (liberty) sense entails the entitlement to be free from coerced gamete donations. It could also be understood to include, for women, entitlement to be free from coerced heterosexual intercourse. The right not to reproduce in the strong (welfare) sense is the right of access to services such as abortion and contraception that enable the individual to avoid procreation. I would argue that failure to recognize both of these forms of the right not to reproduce would result in a kind of reproductive slavery, especially for women, by giving foetuses the right to the use of women's bodies; hence recognition of the right not to reproduce in both its forms is justified.

Chapter 1 also points out some ways in which feminist moral concepts have been co-opted to defend non-feminist claims. For example, some opponents of abortion have argued that the practice of abortion itself is sexist in so far as it allegedly represents a failure to recognize women's unique reproductive needs and capacities. Such a claim demonstrates the necessity of greater clarity about the moral and metaphysical status of pregnancy. The second chapter in this volume, 'Mother/Foetus/State Conflicts', assesses abortion policies in light of various understandings of the relationship between the pregnant woman and her foetus. It argues that law-makers should re-evaluate the adversarial and individualistic perspective that pits the foetus against the pregnant woman and makes the physician or the father or the state the foetus's advocate. Forcing women to continue unwanted pregnancies does not protect foetuses. Incarcerating allegedly abusive pregnant women does not protect foetuses. Subjecting women to foetal surgery or Caesarean sections against their will does not protect foetuses. Instead, protecting and caring for foetuses means supporting and respecting pregnant women, through adequate housing, nutrition, education, and medical care, and freedom from physical and emotional violence and abuse. It also means taking real steps to ensure that no woman is subjected to sexual coercion. In other words, as feminist ethicists have frequently pointed out, the assertion and protection of rights for individuals must be accompanied by recognition and acceptance of responsibilities—in this context, responsibilities belonging to the state, to the health-care and educational systems, to physicians, and to individuals.

Conservatives tend to see the embryo/foetus as a person—or at least a 'potential person'—within a person. From this point of view, the foetus is a kind of tenant, which it is legitimate for physicians to treat along with—or even in spite of—its 'house', the pregnant woman. However, such a perspective sets the stage for a clash of rights between pregnant woman and foetus, and may encourage caregivers to see pregnant women as being in an adversarial relationship with their foetuses.

One alternative to the conservative perspective is to construe the foetus as an organ or organ-like entity within the woman, as simply 'part of her body'. Some liberal and feminist defenders of abortion have taken this view, but it seems ontologically over-simplified. After all, bodily organs do not become persons, and are incapable of surviving outside their body of origin. Moreover, it is not unreasonable to suppose that a woman may have some moral responsibilities to her foetus, at least in a wanted pregnancy, but we do not usually suppose that people have or can have responsibilities to specific bodily organs.

Barbara Katz Rothman believes that the foetus is 'part of the mother's body as long as it is in her body',[9] and proposes that it should be seen as whatever the pregnant woman takes it to be. For some women the foetus is a much-loved baby; for others it is something to be aborted. Thus abortion and contraception are equal, and equally acceptable: both are ways of avoiding 'an unwanted relationship', and 'until we [women] feel we've made a baby, an abortion is stopping a baby from happening, not killing one'.[10] However, one problem with this relativistic approach is that, while it seems faithful to individual variations in the experience of pregnant women, it is not clear how useful it is for social-policy formation on issues such as abortion, foetal-tissue transplantation, and embryo experimentation, which seem to require global generalizations about the status of embryos and foetuses and what may or may not be done with them.

In light of difficulties such as these, the next three chapters in this collection concentrate on issues relating to the handling and disposition of embryos and foetuses, and indirectly attempt to explore some problems relating to the status of the embryo/foetus and its relationship to the pregnant woman.[11] Chapter 3, 'Selective Termination in Pregnancy and Women's Reproductive Autonomy', examines a new and growing technology, related to abortion, that aims at eliminating one or more foetuses within a multiple pregnancy while maintaining the life of the remaining foetus(es). The chapter originated in response to a claim, made by a member of the media, that women in North America are 'forcing' their physicians to provide this seemingly drastic treatment in response to the increasing numbers of multiple pregnancies generated through the use of fertility drugs and *in vitro* fertilization. Puzzled by the notion that women could compel doctors to do anything, I set out to investigate the social context of this practice: the ways in which the cultural creation both of infertility and of the 'need' for a baby generates the 'demand' for selective termination in multiple pregnancy.

The next chapter, 'Biological Mothers and the Disposition of Foetuses after Abortion' considers the difficult problem of what to do about and with those few late-gestational foetuses that survive abortion. Who is entitled to make the decision about how they should be treated, and what should be the basis for that decision? I wrote the paper in order to rethink my earlier claim, made in *Ethics and Human Reproduction*, that while foetuses have no right to occupancy of pregnant women's bodies, pregnant women have no right to kill their foetuses. Hence, I thought, if a woman is seeking an abortion, especially a late abortion, then she is morally obligated to use whatever abortion technique is least likely to damage the foetus and most likely to permit its survival. This claim was criticized by a number of feminist reviewers, who argued that my view was too concerned with the preservation of the foetus, and insufficiently concerned with the health and self-determination of the pregnant woman, who could in fact be trusted to make the right decisions about disposition of the foetus. After considering a number of arguments against saving the foetus, I agree in 'Biological Mothers' that the parturient woman is the appropriate person to make decisions about the disposition of the foetus/infant that survives abortion,[12] but argue that such decisions should attend to the interests of the foetus/infant, not the original reproductive plans of the biological parents.

'Frozen Embryos and "Fathers' Rights": Parenthood and Decision-Making in the Cryopreservation of Embryos' examines the moral, cultural, and policy implications of the Davis v. Davis case in the United States, in which a woman and a man, Mary Sue Davis and her former husband Junior Davis, battled over the fate of seven cryopreserved embryos produced during their marriage through *in vitro* fertilization using their gametes. *Prima facie*, this case is troubling for those who would defend a right not reproduce: Junior Davis argued on appeal that he had been denied his reproductive rights by an earlier judicial decision that gave the embryos to Mary Sue Davis for implantation in her uterus. While Mary Sue Davis argued that the embryos were 'children' that she wanted to nurture and develop, Junior Davis argued that he should not be forced to reproduce against his will. Yet this claim turns out to be another example of the enlistment of a feminist value, this time the appeal to a right not to reproduce, within a non-feminist context. 'Frozen Embryos and "Fathers' Rights"' explores the implications of the Davises' claims for our views about embryos and children, and argues in favour of giving the embryos to the woman—primarily because of the considerable effort, pain, and risk that generating more embryos would otherwise cause her.

The next two chapters investigate the moral dimensions of contract motherhood. 'The Co-optation of Feminist Values in Defence of Reproductive Engineering: A Case Study' examines and responds to a specific set of arguments, promulgated by the philosopher Harriet Baber, that defend contract motherhood arrangements. Baber's paper provides a third example of the misuse of feminist ethics: the language and concepts of progressive social change are exploited in defence of a practice, contract motherhood, that manifestly exploits the women involved and treats their infants as saleable products. Several of the arguments introduced in this chapter are elaborated in the following one, 'The Case Against the Legalization of Contract Motherhood', which argues against the state institutionalization of contract motherhood arrangements. It advances two main reasons for such a policy: first, that most instances of contract motherhood are exploitive of the women who are hired to gestate infants; and second, that all cases of contract motherhood involve the surrender or sale of the infant in return for payments of various sorts, and hence constitute a form of ownership of the child that amounts to slavery.

The eighth chapter in this collection, 'Access to *In Vitro* Fertilization: Costs, Care, and Consent', considers some pros and cons of continuing to offer IVF as part of 'treatment' for infertility, and explores the conditions under which IVF should be made available, in order to maximize women's ability to make an informed choice to use this technology. This assessment of IVF has a ready connection to general issues of health-care resource allocation.

The value of *in vitro* fertilization cannot be assessed in isolation, but must be considered in relation to preventive programs designed to reduce the incidence of infertility, regulatory regimes that minimize threats to reproductive health in the workplace, and educational programs intended to foster responsible sexual behaviour.

The last chapter, 'Reproductive Engineering and Genealogy', was originally written in response to an intriguing invitation from the Ontario Genealogical Society to assess the problems and prospects that technologies such as IVF and practices such as contract motherhood might create for future historians. Presenting an overview of reproductive technologies and the problems they raise, the chapter speculates about the ways in which new and developing reproductive practices and technologies may affect future historical research in genealogy, the systematic investigation of lineage and pedigree.

With the exception of philosophical discussions of abortion, most topics in reproductive ethics have been discussed for little more than a decade. During that time, new reproductive technologies and practices have proliferated, and scholarly and popular debates about them have burgeoned. Our apparently private, personal procreative behaviour continues to generate ethical and political questions—questions that are ultimately about what kinds of persons we should be and become, and what sort of society we should create. I hope this book will contribute to the ongoing exploration of these issues.

Notes

[1]Caroline Whitbeck, 'The Moral Implications of Regarding Women As People: New Perspectives on Pregnancy and Personhood', in William B. Bondeson, H. Tristram Engelhardt, Jr, Stuart F. Spicker, and Daniel H. Winship, eds, *Abortion and the Status of the Fetus* (Boston: D. Reidel, 1984), 264.

[2]This book does not use the term 'pre-embryo', a recently coined word used to denote the earliest stages of development of the fertilized egg, up until the formation of the 'primitive streak' at day fourteen, before the individuality and continuity of the embryo have been established. (See, for example, Peter Singer, Helga Kuhse, Stephen Buckle, Karen

Dawson, and Pascal Kasimba, *Embryo Experimentation* [Cambridge: Cambridge University Press, 1990], where the term 'pre-embryo' is used extensively.) It has been argued by conservatives and feminists alike that the effect of the creation and promotion of this term is to minimize the importance of the human embryo and thus provide a covert justification of embryo experimentation, a practice that both conservatives and feminists usually find morally problematic, although for different reasons. While the conservative believes in full moral status for human entities from the time of conception, and therefore does not want to concede any term that would minimize its significance, feminists have been sceptical of the uses of embryo research, and concerned about the harms to women's bodies that producing a sufficient supply of embryos may cause.

[3]For two reasons the use of the term 'donor insemination' is preferable to 'artificial insemination by donor' (AID). First, it seems worthwhile to avoid misunderstandings that might be caused by the accidental confusion of the acronym AID with AIDS (acquired immunodeficiency syndrome). Second, the use of the word 'artificial' has been challenged by some feminist theorists on the grounds that we should be suspicious of the implicit contrast between 'artificial' and 'natural': all forms of reproductive behaviour are 'artificial' in so far as they reflect human choices, customs, and values, rather than the mere rote enactment of instinctual behaviour. And many lesbians, who are among the users of donor insemination, have been critical of the implication that their way of conceiving a child is less 'natural' than heterosexual intercourse.

Notice, also, that the use of the word 'donor' masks the fact that sperm providers are usually paid a substantial sum of money for their few minutes' 'work', and implies, questionably, that their motivation is purely altruistic. For this reason, under most circumstances 'vendor' seems to be more accurate term, although 'vendor insemination' sounds peculiar. Some feminists have therefore advocated use of the term 'alternative insemination'.

[4]Congregation for the Doctrine of the Faith, 'Instruction on Respect for Human Life in Its Origin and on the Dignity of Procreation: Replies to Certain Questions of the Day'. In Richard T. Hull, ed., *Ethical Issues in the New Reproductive Technologies* (Belmont, CA: Wadsworth, 1990), 21-39.

[5]One example is 'Ethical Considerations of the New Reproductive Tech-

nologies', by the Ethics Committee of the American Fertility Society, in Hull, ed., *Ethical Issues*, 40-8.

[6]See Lori Andrews, 'Position Paper: Alternative Modes of Reproduction', in Sherrill Cohen and Nadine Taub, eds, *Reproductive Laws for the 1990s* (Clifton, NJ: Humana Press, 1989), 361-403; and Carmel Shalev, *Birth Power: The Case for Surrogacy* (New Haven: Yale University Press, 1989).

[7]See Gena Corea, *The Mother Machine: Reproductive Technologies From Artificial Insemination to Artificial Wombs* (New York: Harper and Row, 1985), and Barbara Katz Rothman, *Recreating Motherhood: Ideology and Technology in a Patriarchal Society* (New York: W.W. Norton, 1989).

[8]*Ethics and Human Reproduction: A Feminist Analysis* (Boston: Allen and Unwin, 1987).

[9]Rothman, *Recreating Motherhood*, 258.

[10]Ibid., 107 and 123.

[11]In most of the papers here I have chosen not to call the pregnant woman a 'mother'. While some pregnant women regard themselves as mothers, others do not. The use of the term—except to describe a woman with existing offspring—seems to beg some crucial questions about the moral status of the foetus and its relationship to the pregnant woman, and to imply that foetuses are tiny undeveloped children.

[12]The argument that the birthing woman is the appropriate person to decide about the disposition of the foetus clearly also applies to decision-making with respect to the use of foetal tissue for transplantation and experimentation.

Chapter 1

♦ ♦ ♦

Reflections on
Reproductive Rights in Canada

The concept of 'reproductive rights' plays a central role in discussions of issues relating to women's reproductive health. While the concept of rights in general does not by any means constitute all that is important in ethical discourse, and the concept of reproductive rights in particular does not exhaust all that is significant in the moral evaluation of reproductive issues, the idea of reproductive rights is, at this point at least, indispensable to a complete discussion of reproductive ethics and social policy.

Barbara Katz Rothman has expressed reservations about the use of the term 'reproduction', arguing that we do not literally produce babies or reproduce ourselves.[1] While I agree with this observation, I shall continue to use the phrase 'reproductive right' because I specifically want to explore the strengths and the ambiguities of that phrase. In particular, this chapter will analyse and evaluate the ways in which this notion of reproductive right has been given a unique legal and moral expression within Canadian society during the last two decades, manifesting itself within the struggle for abortion, the debates about midwifery and the place of birth, and the introduction of social practices relating to new reproductive technologies such as *in vitro* fertilization.

Originally, the concept of reproductive rights seemed to find a natural home within the abortion debate, but it is now being extended to discussions of new reproductive technologies. What is extraordinary is that the idea of reproductive rights is used not only by feminists concerned with promoting women's well-being and ending oppression, but also by non-feminists, whose agenda usually includes preservation of the traditional family, extension of male sexual and reproductive entitlements, and enforcement of a 'pro-life' morality that sees women's bodies as instruments for the production of babies. In other words, while the idea of reproductive rights is both useful and central to the advancement of women's reproductive freedom, there are also ways in which it can be and is being used against women's best interests. For this reason feminists need to think carefully about what they mean when making claims for reproductive rights.

◆ ◆ ◆
The Right Not to Reproduce

The term 'reproductive right' has more than one meaning.[2] It is necessary, first, to distinguish between the right to reproduce and the right *not* to reproduce. The two are sometimes mistakenly conflated as, for example, when Justice Bertha Wilson referred in her Supreme Court decision on the Morgentaler case to '[t]he right to reproduce or not to reproduce which is in issue in this case'.[3] The right not to reproduce is the entitlement not to be compelled to beget or bear children against one's will: the entitlement not to have to engage in forced reproductive labour. To say that women have a right not to reproduce implies that they have no obligation to reproduce. In its weak (liberty) sense it is the entitlement not to be compelled to donate gametes (eggs or sperm) or embryos against one's will. The right not to reproduce in the strong (welfare) sense is the right of access to services like abortion and contraception that enable women to avoid procreation. Women do not owe their reproductive products or labour to any person or institution, including male partners or the state.

Recognition of the right not to reproduce has been slower to develop in Canada than in the United States. Full exercise of such a right requires, among other protections, access to safe and effective contraception and abortion services. In other words, the moral entitlement to abortion access follows from the broader right not to reproduce. In 1969, stipulations were introduced into the Criminal Code that provided for the possibility, under certain carefully specified conditions, of therapeutic exceptions to the general law that using any means or permitting any means to be used for the purpose of procuring a miscarriage on a female person was an indictable offence. According to the then new Section 251 of the Code, such a therapeutic abortion had to be performed by a qualified medical practitioner who was not a member of a therapeutic abortion committee; in an accredited or approved hospital; and only after a decision by the hospital's therapeutic abortion committee, consisting of at least three members, each of them a qualified medical practitioner, that the continuation of the pregnancy would, or would be likely to, endanger the life or health of the pregnant woman.

In the ensuing years, variations in interpretation of and conformity to Section 251 gave rise in many cases to outright injustice. In some areas hospitals could not obtain accreditation or approval; in some hospitals there were not sufficient doctors to constitute a therapeutic abortion committee. Hospitals had no legal obligation to set up such a committee; and those committees that did exist had no obligation to meet. Some hospitals imposed quotas on the numbers of abortions performed, or limitations on patient eligibility based on place of residence. Women had no right to appear before the committees to present their case, and interpretation of the phrase 'life or health' of the pregnant woman was left entirely to committee members. Interpretations varied enormously from one committee to another, and some even introduced extraneous considerations such as marital status of the applicant, consent of her spouse, or number of previous abortions.[4]

Section 251 of the Criminal Code placed severe constraints on Canadian women's right not to reproduce. In effect, it said that some women—directly, those whose abortion requests were rejected by therapeutic abortion committees, and indirectly, those who had no opportunity to bring their request to such a committee—had a legal obligation to procreate; it sentenced them to forced reproductive labour. This surely was a major violation of what the Charter of Rights and Freedoms now refers to as 'security of the person'. In the words of Supreme Court Judge Jean Beetz in the Morgentaler decision, 'A pregnant woman's person cannot be said to be secure if, when her life or health is in danger, she is faced with a rule of criminal law which precludes her from obtaining effective and timely medical treatment.'[5]

The so-called 'pro-life' movement in North America places heavy emphasis upon what is alleged to be the foetus's right to life. Even without specifically recognizing such a right, however—indeed, without making any direct references at all to the foetus or its physical condition—Section 251 implicitly attributed to the foetus a right to the use and occupancy of the woman's uterus. In effect, the woman's body was regarded simply as a container, with various utilities, that the foetus happened to need for nine months. Indeed, foetuses are the only group of entities that have been given entitlement under Canadian law to the medical use of the bodies of adult persons.

Section 251 also helped to perpetuate a right of access to women's reproductive labour that potentially benefited both individual men and the state. In the words of Justice Bertha Wilson, Section 251 of the Criminal Code 'assert[ed] that the woman's capacity to reproduce is not to be subject to her own control. It is to be subject to the control of the state.... She is truly being treated as a means—a means to an end which she does not desire but over which she has no control.'[6]

Fortunately, in January 1988 the Supreme Court removed these Criminal Code impediments to women's access to abortion.[7] No longer is a woman seeking an abortion required to obtain the approval of a therapeutic abortion committee. According to the judicial decision, 'Forcing a woman, by threat of criminal sanction, to carry a foetus to term unless she meets certain criteria unrelated to her own priorities and aspirations, is a profound interference with a woman's body and thus a violation of security of the person.'[8]

But this decision has by no means permanently removed a significant danger to women's right not to reproduce. Like women in the United States,[9] Canadian women cannot assume that access to abortion will remain indefinitely protected. Potential threats to abortion access can be found in at least three areas.

First, there is a danger that the recent expression of concerns about the reasons for abortion and abortion-related procedures may lead to renewed limitations on access to abortion. One example is the growing media discussion of abortion for the purpose of sex selection.[10] Another is the demand for regulation of and limitations on so-called selective termination in pregnancy (discussed in detail in Chapter 3 below). This procedure, performed in cases of multiple pregnancy in order to reduce the number of foetuses in the uterus, usually involves the injection of potassium chloride into the thorax of one or more of the foetuses to stop the heart. The 'terminated' foetus is reabsorbed into the woman's body, without further need for surgery.[11] Recent media news reports have quoted Canadian ethicists and physicians as challenging the justification of the procedure and calling for limitations on the number of foetuses to be terminated.[12]

A second reason for concern about potential threats to the right not to reproduce is the persistence of the claim that there is a need for protection of foetal life and alleged foetal rights. In March 1989 the Supreme Court of Canada dismissed Joseph Borowski's argument that Section 251 of the Criminal Code contravened the life, security, and equality rights of the foetus, as a 'person' protected by sections 7 and 15 of the Canadian Charter of Rights and Freedoms. In the absence of a law governing abortion, the Court found the appeal to be moot, and stated that the appellant no longer had standing to pursue the appeal. It added:

In a legislative context any rights of the foetus could be considered or at least balanced against the rights of women guaranteed by s. 7.... A pronouncement in favour of the appellant's position that a foetus is protected by s. 7 from the date of conception would decide the issue out of its proper context. Doctors and hospitals would be left to speculate as to how to apply such a ruling consistently with a woman's rights under s. 7.[13]

Nevertheless, it is not impossible that a future government could use the decision in the Borowski case as part of the rationale for reintroducing a law to recriminalize some abortions. Indeed, even the 1988 decision striking down the existing abortion law left open the very real possibility that legal steps might be taken to protect so-called foetal rights. For example, Justice Wilson stated that 'Section 1 of the Charter authorizes reasonable limits to be put upon the woman's right having regard to the fact of the developing foetus within her body.' She added:

A developmental view of the foetus ... supports a permissive approach to abortion in the early stages of pregnancy and a restrictive approach in the later stages.... [The woman's] reasons for having an abortion would ... be the proper subject of inquiry at the later stages of her pregnancy when the state's compelling interest in the protection of the foetus would justify it in prescribing conditions. The precise point in the development of the foetus at which the state's interest in its protection becomes 'compelling' I leave to the informed judgment of the legislature which is in a position to receive guidance on the subject from all the relevant disciplines. It seems to me, however, that it might fall somewhere in the second trimester.[14]

Similarly, Chief Justice Brian Dickson said, 'State protection of foetal interests may well be deserving of constitutional recognition under s. 1.'[15] And Justice Beetz stated:

[A] rule that would require a higher degree of danger to health in the latter months of pregnancy, as opposed to the early months, for an abortion to be lawful, could possibly achieve a proportionality which would be acceptable under s. 1 of the Charter.... Parliament is justified in requiring a reliable, independent and medically sound opinion in order to protect the state interest in the foetus.... [T]here would be a point in time at which the state interest in the foetus would become compelling. From this point in time, Parliament would be entitled to limit abortions to those required by therapeutic reasons and therefore require an independent opinion as to the health exception.... I am of the view that the protection of the foetus is and ... always has been, a valid objective in Canadian criminal law.... I think s. 1 of the Charter authorizes reasonable limits to be put on a woman's right having regard to the state interest in the protection of the foetus.[16]

More recently it has been claimed that without protection of foetal rights there is nothing to prevent abortion for purposes of sex selection, harm to foetuses by the use of dangerous drugs and by third-party attacks, and the buying and selling of foetuses and foetal parts.[17] The Law Reform Commission of Canada's Working Paper entitled 'Crimes Against the Foetus' expresses serious concern about dangers to the foetus, and proposes a new category of criminal offence, 'Foetal Destruction or Harm'.[18] Thus the stage is set for a potential continuation of major conflict between women's right not to reproduce and the alleged rights of the foetus. Unfortunately, as the Working Paper makes clear, recognition of foetal rights would almost inevitably mean the judicial recognition, via the recriminalization of abortion,[19] of the foetus's alleged right to occupancy and use of a woman's body, and concomitant limitations on women's autonomy and self-determination.

A third reason for concern about threats to Canadian women's right not to reproduce is the growing use by anti-abortion groups of not only non-violent civil disobedience but also active interference in the operations of abortion clinics.[20] In addition, 'pro-life' leaders such as Joseph Borowski have threatened the use of violence in defence of their cause:

> The war goes on. There is no end to this fight, and certainly no compromise. They [pro-choice] are the enemy. It's a war. Our side has one advantage. We pray. They don't.... I'm glad I did not come to Ottawa for the [Morgentaler Supreme Court] decision. I probably would have gone into the court and punched the judges in the nose.... I'm a non-violent man, and I don't believe in violence but if the seven judges or whoever were here right now, I would have great difficulty restraining myself from punching them in the mouth.[21]

In the face of these potential threats to the right not to reproduce, feminist research and activism must insist that there is no need for the recriminalization of abortion, including late abortion. No woman deliberately sets out to kill a highly developed foetus, and abortion is not sought by women for its own sake. Rather, in the words of philosopher Caroline Whitbeck, it is often a 'grim option'.[22] While abortions late in pregnancy may seem particularly problematic, they are usually requested for one of the following reasons. In some instances it was impossible for the woman to obtain the abortion earlier, because convoluted legal procedures delayed its approval. Cases such as these can be avoided by making very early abortions easily available and accessible. In other instances, a late abortion is sought either because prenatal testing reveals a foetal condition that is or is perceived as being severely disabling or even life-threatening, or because the woman's own life or health is endangered.[23] There is, therefore, insufficient justification for the introduction of legislation to protect the late-term foetus from the pregnant woman, or indeed for any new Criminal Code limitations on abortion.

Moreover, the entitlement to choose how many foetuses to gestate, and of what sort, should be seen as a part of the right not to reproduce. The protection of this choice is essential within a cultural context where mothering receives little social support, people with disabilities are subject to bias and stigmatization, and raising several infants simultaneously represents a personal and financial challenge of heroic dimensions. (It is also significant that the upsurge in the incidence of multiple pregnancies has been generated by the administration of fertility drugs and by the use of *in vitro* fertilization and the technology of gamete intrafallopian transfer, or GIFT.) There is no more reason to demand that a woman gestate a certain number of foetuses or type of foetus than there is to demand that she gestate a given foetus or foetuses. To set such requirements is to accord those foetuses an unjustified right of occupancy of the woman's uterus.

If protection of the foetus seems to be a worthwhile and neglected social goal, then what is needed is greater protection of the pregnant woman herself. Even, surely, on the basis of the non-feminist and implausible assumption of an adversarial relationship between pregnant woman and foetus, the reinstallation of physicians as 'body police' enforcing foetal rights[24] is unlikely to improve the behaviour of pregnant women towards their foetuses. Furthermore, in order to prevent the commodification of foetuses and foetal parts, and other undesirable uses of the foetus, there is no more need to assign personhood or rights to the foetus than there is to assign personhood or rights to blood or body parts. Instead of using that blunt instrument the criminal law in *post hoc* fashion to attempt to manage undesirable reproductive practices, use can be made of existing regulations governing health care and the utilization of human tissues. Ultimately, of course, it will be necessary to minimize and finally eliminate the powerful underlying conditions of oppression that generate such practices as foetal commodification and sex selection.

♦ ♦ ♦

The Right to Reproduce

The right not to reproduce is distinct from the right to reproduce. In other words, it is independent, neither implying a right to reproduce nor following from such a right.

The right to reproduce has two senses, which may be called the weak (or liberty) sense and the strong (or welfare) sense. The weak sense of the right to reproduce is the entitlement not to be interfered with in reproduction, or prevented from reproducing. It would imply an obligation on the part of the state not to inhibit or limit reproductive liberty through, for example, racist marriage laws, forced sterilization,[25] or coercive birth-control programs.

In this weak sense, the right to reproduce is also compromised by restrictions on the place of birth and on birth attendants, and by court-ordered Caesarean sections for competent but unwilling pregnant women. Like the United States, Canada has a history of the gradual medicalization of birth. Midwives have been replaced by physicians, from general practitioners to obstetricians; hospitals have replaced the home; and medical innovations from foetal monitoring, amniotomy (the premature puncturing of the amniotic sac); and forceps deliveries to anaesthesia and Caesarean sections have made birthing into a health crisis. Without the genuine freedom to choose home birth, to be attended by midwives, and to avoid obstetrical technology, women's right to reproduce in the weak sense is seriously compromised.

In its strong sense, the right to reproduce would be the right to receive all necessary assistance to reproduce. It would imply entitlement to access to any and all available forms of reproductive products, technologies, and labour, including the gametes of other women and men, the gestational services of women, and the full range of procreative techniques including *in vitro* fertilization, gamete intrafallopian transfer, uterine lavage (a process in which one woman is inseminated with sperm, her uterus is flushed with fluid, and the embryo is retrieved for implantation in another woman's uterus), embryo freezing, and sex preselection.

Non-feminist writers such as American legal theorist John A. Robertson defend the right to reproduce in the strong sense by claiming that it is simply an extension of the right to reproduce in the weak sense. As he puts it, 'the right of the married couple to reproduce noncoitally' and 'the right to reproduce noncoitally with the assistance of donors and surrogates' both follow from 'constitutional acceptance of a married couple's right to reproduce coitally'.[26] (Robertson's heterosexist bias is not much mitigated by his later concession that there is 'a very strong argument for unmarried persons, either single or as couples, also having a positive right to reproduce'.[27]) Robertson believes that these rights entitle married couples certainly, and perhaps single persons, to 'create, store, transfer, donate and possibly even manipulate extra-corporeal embryos'; and 'to contract for eggs, sperm, embryos, or surrogates'. They would also, he thinks, justify compelling a contract mother to hand over a child to its purchasers, even against her will.[28]

In addition, American attorney Lori B. Andrews argues that the right to reproduce in the strong sense is probably founded upon the right to marital privacy, which, she claims, protects the full range of married people's choices about both sexual and reproductive behaviour.[29] Hence some feminists may want to claim the right to reproduce in the strong sense both because of arguments such as those of Robertson and Andrews, and because of a fear that otherwise access to reproductive technologies such as IVF may be treated by the state as a privilege—one to be gained only through possession of the requisite social criteria, such as being heterosexual and married.

Nevertheless, the legitimacy and justification of this right to reproduce are questionable. To recognize it would be to shift the burden of proof onto those who have doubts about the morality of technologies such as IVF and practices such as contract motherhood. For it suggests that a child is somehow owed to each of us, as individuals or as members of a couple, and that it is indefensible for society to fail to provide all possible means for obtaining one. Thus it might be used, as Robertson advocates, to imply an entitlement to hire contract mothers, to obtain other women's eggs, and to make use of donor insemination and uterine lavage of another woman, all in order to maximize the chances of reproducing.[30] In other words, recognition of the right to reproduce in the strong sense would create an active right of access to women's bodies and in particular to our reproductive labour and products. For example, it would condone the hiring of contract mothers, and force the latter to surrender their infants after birth. And it might be used to found a claim to certain kinds of children—for example, children of a desired sex, appearance, or intelligence.

Exercise of the alleged right to reproduce in this strong sense could potentially require violation of some women's right not to reproduce. There is already good evidence, in both the United States and Great Britain, that eggs and ovarian tissue have been taken from some women without their knowledge or informed consent.[31] It is not difficult to imagine that recognizing a strong right to reproduce could require either a similar theft of eggs or embryos from some women, if none can be found to offer them willingly, or a commercial inducement to sell these products. Such a right could be used as a basis for requiring fertile people to 'donate' gametes and embryos. Even if some people would willingly donate gametes, there is no *right* on the part of the infertile that would entitle them to demand such donations.

The feminist language of reproductive rights is illegitimately co-opted when it is used to defend an alleged right to become or to hire a contract mother, to buy or to sell eggs, embryos, or babies, or to select or preselect the sex of one's offspring. There can be no genuine entitlement to women's reproductive labour, nor to buying or otherwise obtaining human infants. Contract motherhood entails a type of slave trade in infants, and it commits women to a modern form of indentured servitude.

It is to be hoped that Canada will choose neither the legalization of contract motherhood nor the criminalization of contract mothers. We should opt instead to reduce the potential motivation for such contracts by making them unenforceable and by rendering criminal both the operation of contract motherhood agencies and the actions of professionals who participate in such arrangements. It is important for Canadian social policy to resist the incursion of US-style commercialization of reproduction and reproductive entrepreneurialism, the most likely victims of which would be poor women and women of colour (see Chapter 7 below).

At the same time, there is no need to treat procedures like *in vitro* fertilization as privileges to which access may be limited on arbitrary and unfair grounds—grounds such as marital status, sexual orientation, putative stability or parenting potential, or economic level. While I cannot wholeheartedly support and endorse highly ineffective, costly, and painful procedures such as IVF, I also cannot endorse the call by some feminists for a total ban on the procedure. State provision and financing of IVF is different from state provision of contract motherhood. Many compelling reasons—primary among them being the sale of babies and the exploitation of women's reproductive labour—militate against state recognition of contract motherhood through legalization or financial support. These two reasons are not present, or not inevitably present, in the case of IVF.

Rather, without asserting a strong right to all possible reproductive assistance, we can critically examine the artificial barriers, such as marital status, sexual orientation, and ability to pay, that get in the way of women's fair access to reproductive technologies. We can also provide protections for women entering and participating in infertility treatment programs. This would require ensuring that applicants make a genuinely informed choice, in full knowledge of the short- and long-term risks, possible benefits, chances of success and failure, alternative approaches and treatments, and perhaps even the pronatalist social pressures to procreate. If IVF seems to be a valuable medical service (and that view is still debatable) then it deserves to be made available, like other medical services, through medicare, as is now the case in Ontario. It would also be important to ensure thorough screening of egg and sperm donors; to maintain adequate records that will make it possible to track the long-term effects of IVF on women and their offspring; and to ensure that any women who provide eggs for the program have genuinely chosen to do so. Finally, in the long run, feminists should be thinking about whether it is possible to incorporate high-tech infertility treatments such as IVF into women-centred and women-controlled reproductive health centres.

The approach sketched here avoids two tendencies that I believe are undesirable: on the one hand, treating access to reproductive technology as a privilege to be earned through the possession of certain personal, social, sexual, and/or financial characteristics; and on the other hand, a kind of feminist maternalism that seeks, in the best interests of women, to terminate IVF research and treatment.[32] While many feminists have stressed both the social construction of the desire for motherhood and the dangers and ineffectiveness of *in vitro* fertilization,[33] it is surely dangerous for feminists to claim to understand better than infertile women themselves the origins and significance of their desire for children. It is not the role of feminist research and action to protect women from what is interpreted as their own 'false consciousness'. Instead, we should assume that when women are provided with full information about the possibilities they will be empowered to make reproductive decisions that will genuinely benefit themselves and their children.[34]

♦ ♦ ♦
Conclusion

The two themes that have structured the struggle over reproductive rights in Canada appear to have little in common: on the one hand, access to various reproductive services and technologies; on the other, the use and exploitation of women's bodies for reproductive purposes.

Nevertheless, the goals of access to reproductive technologies and access to women's bodies come together within conservative discourse on reproduction. Pronatalist, pro-family, and in favour of traditional roles for women, this discourse is also classist, racist, ableist, and heterosexist. The oppressive nature of this discourse is often disguised by the co-optation of feminist language and concepts. In recent lectures and papers, for example, members of the anti-abortion movement have claimed that there is 'sexism' in the pro-choice movement,[35] which is alleged to favour the sexual agenda of men over the reproductive needs of women, and have depicted the foetus as a member of a maligned minority group. The same voices that want to ban abortion because women engage in sexual intercourse 'by choice' also want to compel contract mothers to sell their babies because they enter the contracts 'by choice'. Meanwhile, non-feminists are proclaiming that new reproductive technologies actually promote women's autonomy and reproductive choice.[36]

Paradoxically, however, both the non-existence and the existence of certain reproductive 'choices' or alternatives can be coercive. While lack of access to contraception or abortion clearly violates reproductive choice by failing to respect the right not to reproduce, conversely practices such as contract motherhood and the sale of gametes and embryos have the potential to violate reproductive choice. In other words, respecting a right to reproduce in the strong sense for some may violate the right *not* to reproduce of others.

Feminists want to preserve and enhance access to the reproductive services and technologies that benefit women while preventing further encroachments on access to women's bodies, whether by the state or by individuals. The way to do this is by insisting upon both women's right not to reproduce and our right to reproduce in the weak sense, and also by developing a critical analysis of the ways in which the right to reproduce in the strong sense is now being exercised.

Notes

[1] Barbara Katz Rothman, remarks at a conference on 'Legal and Ethical Aspects of Human Reproduction', Canadian Institute of Law and Medicine, Toronto, Dec. 1989.

[2] Christine Overall, *Ethics and Human Reproduction: A Feminist Analysis* (Boston: Allen and Unwin, 1987), Chapter 8.

[3] R. v. Morgentaler [1988] 1 SCR 30, 172.

[4] Ibid., 56-61, 64-73.

[5] Ibid., 90.

[6] Ibid., 173.

[7] Ignoring the Criminal Code provisions on abortion, community health clinics in Quebec had already been providing abortions for more than a decade before the Supreme Court decision.

[8] R. v. Morgentaler [1988] 1 SCR 32.

[9]See George J. Annas, 'Webster and the Politics of Abortion', *Hastings Center Report* 19 (March/April 1989): 36-8. In July 1989 the United States Supreme Court ruled that the state of Missouri had the right to ban the performance of abortions by public hospitals and public employees. It is anticipated that the court may in future uphold additional state-imposed restrictions on abortion services, seriously limiting access to abortion, particularly for poor women. See *Webster v. Reproductive Health Services*, 109 S. Ct. 3040 (1989).

[10]For example, Brenda Large's editorial 'If Sex-Based Abortions are Wrong, So Are All', *Kingston Whig-Standard* (11 Feb. 1989).

[11]Mark I. Evans, John C. Fletcher, Evan E. Zador, Burritt W. Newton, Mary Helen Quigg, and Curtis D. Struyk, 'Selective First-Trimester Termination in Octuplet and Quadruplet Pregnancies: Clinical and Ethical Issues', *Obstetrics and Gynecology* 71, 3, pt. 1 (March 1988): 291. For an extensive discussion of the ethics and social context of this procedure, see 'Selective Termination in Pregnancy and Women's Reproductive Autonomy', chapter 3 below.

[12]Dorothy Lipovenko, 'Infertility Technology Forces People to Make Life and Death Choices', *The Globe and Mail* (21 Jan. 1989): A4; 'Multiple Pregnancies Create Moral Dilemma', *Kingston Whig-Standard* (21 Jan. 1989): 3.

[13]Borowski v. Canada (Attorney General) [1989] 1 SCR. 342.

[14]R. v. Morgentaler, 181, 183.

[15]Ibid., 76.

[16]Ibid., 82-3, 110, 113, 124.

[17]Neil Reynolds, 'Fetal Status Cannot Depend on Momentary Opinion', *Kingston Whig-Standard* (16 March 1989): 6.

[18]Law Reform Commission of Canada, Working Paper 58, 'Crimes Against the Foetus' (Ottawa, 1989), 64.

[19]Ibid., 42, 56, and 64.

[20]A. Johnson, 'Clinic Fights to Survive in B.C.', *Rites* (March 1989): 4; and Helen Armstrong, Debi Brock and Jennifer Stephen, '"Operation Rescue" Turns Into Fiasco', *Rites* (March 1989): 5.

[21]Joseph Borowski, quoted in *The Toronto Star* and *The Globe and Mail*, 10 March 1989.

[22]Caroline Whitbeck, 'The Moral Implications of Regarding Women as People: New Perspectives on Pregnancy and Personhood', in William B. Bondeson, H. Tristram Engelhardt, Jr, Stuart F. Spicker, and Daniel H. Winship, eds, *Abortion and the Status of the Foetus* (Boston: Reidel, 1984), 251.

[23]This is argued by Sanda Rodgers in 'The Future of Abortion in Canada', in Christine Overall, ed., *The Future of Human Reproduction* (Toronto: Women's Press, 1989). This argument in no way endorses the use of abortion for supposed eugenic purposes. As disabled women and their allies have pointed out, feminists should be highly critical of the use of reproductive technologies to discriminate among human beings on the basis of their physical or mental condition, or to promote the notion of human perfectibility. See Marcia Saxton, 'Prenatal Screening and Discriminatory Attitudes About Disability', in Elaine Hoffman Baruch, Amadeo F. D'Adamo, Jr, and Joni Seager, eds, *Embryos, Ethics, and Women's Rights: Exploring the New Reproductive Technologies* (New York: Haworth Press, 1988), 217-24, and Ruth Hubbard, 'Eugenics: New Tools, Old Ideas', in ibid., 225-35.

[24]In my view, this is the effect of the proposal of the Law Reform Commission of Canada's 'Crimes Against the Foetus', which would require 'medical authorization' of an abortion by one 'qualified medical practitioner' before foetal viability, and by two such practitioners after the foetus is capable of independent survival (64).

[25]In North America there is a long history of the forced sterilization of native women and women of colour.

[26]John A. Robertson, 'Procreative Liberty, Embryos, and Collaborative Reproduction: A Legal Perspective' in Baruch et al., eds, *Embryos, Ethics and Women's Rights*, 180. Cf. Lori B. Andrews, 'Alternative Modes of Reproduction', in Sherrill Cohen and Nadine Taub, eds, *Reproductive Laws for the 1990s* (Clifton, NJ: Humana Press, 1989), 364.

[27]Robertson, 'Procreative Liberty', 181.

[28]Ibid., 180, 186, and 190.

[29]Lori B. Andrews, *New Conceptions: A Consumer's Guide to the Newest Infertility Treatments* (New York: Ballantyne Books, 1985), 138.

[30]From this point of view, then, IVF with donor gametes is more problematic than IVF in which a woman and a man make use of their own eggs and sperm.

[31]Genoveffa Corea, 'Egg Snatchers', in Rita Arditti, Renate Duelli Klein, and Shelley Minden, eds, *Test-Tube Women: What Future for Motherhood* (London: Pandora Press, 1984), 37-51.

[32]Renate Duelli Klein, quoted in Christine St. Peters, 'Feminist Discourse, Infertility, and the New Reproductive Technologies', *National Women's Studies Association Journal* 1, 3 (Spring 1989): 358.

[33]See, for example, Susan Sherwin, 'Feminist Ethics and In Vitro Fertilization', in Marsha Hanen and Kai Nielsen, eds, *Science, Morality and Feminist Theory* (Calgary: University of Calgary Press, 1987), 265-84.

[34]For a more detailed discussion of the fair provision of IVF, see 'Access to *In Vitro* Fertilization: Costs, Care, and Consent', Chapter 8 below.

[35]See, for example, Diane Marshall and Martha Crean, 'The Human Face of a Woman's Agony', in Ian Gentles, ed., *A Time to Choose Life: Women, Abortion and Human Rights* (Toronto: Stoddart, 1990), 134-42, and Janet Ajzenstat, 'The Sexism of Pro-Choice', Queen's University, Kingston, Ont., 14 March 1989. Ajzenstat is a member of the Department of Political Science, McMaster University.

[36]John Robertson ('Procreative Liberty') claims that '[e]xtra-corporeal conception seems to promote choice, to promote the autonomy of women (and men) in helping them overcome infertility, which for many women (and men) is a very serious problem' (192). It 'makes possible new, partial reproductive roles for women' as 'egg and embryo donors and surrogates' (193).

Chapter 2

◆ ◆ ◆

Mother/Foetus/State Conflicts

Another title for this chapter might be 'Whose Body Is It Anyway?' Whenever the topic of abortion and foetal rights is examined, there is a tendency to lose sight of the fact that what is at stake is the bodily integrity and autonomy of women. A woman's body does not belong to the state; it does not belong to physicians; it does not belong to the woman's husband or partner, or to the father of her children; and, most important, it does not belong to the foetus.

Nevertheless, many people who attempt to defend foetal rights seem to assume that the pregnant woman's body belongs to some or all of these entities. This chapter will examine the underlying assumptions made by defenders of foetal rights—assumptions that, I believe, lack evidence to support them and in some cases are false. Most of what follows is in no way original,[1] but it clarifies some basic distinctions that I think are worth repeating, and is intended to provide an overview of some of the crucial controversies in discussions of foetal rights.

The cornerstone of the foetal-rights advocates' position is the belief that the foetus has the right to life. They seldom defend this claim, perhaps because they do not think it needs defending.

However, we need to see whether there is any evidence for it. Foetal-rights advocates usually rely on the dual claim that the foetus is alive and that it is a human being.[2] It is indeed true that the foetus is alive; it is not inert matter and it is not dead. It is also unquestionable that the foetus is human; it is not canine, feline, equine, or bovine. But what is not uncontroversial is what follows from these minimal facts.

Foetal-rights advocates appear to assume that the fact that the foetus is human and living makes it morally equivalent to a two-year-old child or a twenty-five-year-old woman, and gives it a right to life. Yet many of our other practices—practices that seem eminently reasonable—suggest that there is no such equivalence. For example, the intrauterine device (IUD) operates, in part, by preventing the fertilized ovum from implanting in the wall of the uterus and continuing its development. Yet most people do not regard the IUD as a murder weapon that deprives the fertilized egg of its 'right to life'. In addition, many women suffer unforeseen miscarriages early in their pregnancies. Most people would not advocate a legal investigation to determine whether these women have done something to deprive their foetus of its 'right to life'; no one regards women who have miscarried as possible murderers. Finally, the process of *in vitro* fertilization, by which sperm and ova are combined in a petri dish rather than inside a woman's body, often produces surplus embryos. Most people do not worry that the right to life of these embryos must be protected from potential murderers. If there were a fire in a laboratory, and the technicians failed to save some embryos produced through IVF, most people would not regard that failure as morally serious in the way that a failure to save babies or children would be.

These moral intuitions about the effects of the IUD on embryos, about women who miscarry, and about IVF embryos do not seem unreasonable: eight-cell embryos and three-month-old foetuses do not warrant extraordinary efforts to preserve their lives, and indeed may justifiably be sacrificed when other goals, such as the preservation of reproductive autonomy, are sought. They suggest that the fact that the foetus is living and is human is not in itself sufficient to establish that it has a right to life. In order to establish that a living, human foetus has a right to life, it would be necessary to show that it is a person, in the same way that the two-year-old child or the twenty-five-year-old woman are persons.[3] In light of our other moral intuitions about foetuses, the burden of proof must rest on those who would claim that the foetus is a person with the same moral status as women, men, and children; it cannot simply be taken for granted, or inferred from the fact that the foetus is living and human.

Although foetal-rights advocates stress the alleged right to life of the foetus, they say virtually nothing about what I believe is another, but more covert, set of assumptions: that the foetus has the right to the use of the pregnant woman's body, that that right should be legally protected, and hence that the pregnant woman has an obligation not to abort, and to permit any intervention in her body that is thought medically necessary for the sake of the foetus.

But in a landmark paper originally published in 1971,[4] philosopher Judith Jarvis Thomson pointed out that *even if* we grant the foetus the right to life, it does not follow that the foetus has the right to the use of the pregnant woman's body. For example, imagine that I have a life-threatening disease that can somehow only be alleviated or cured by my making use of your body. Perhaps I need one of your kidneys, or an injection of your bone marrow. Or perhaps I need to be hooked up to your body for weeks or months or years so that I can receive ongoing transfusions of your blood. I have a right to life, we would all agree. But my right to life in no way gives me the right to the use of your body, even if I need it for the continuance of my own life, and even if I am related to you.

The reason for this is simple: for me to claim the right to the use of your body would be an assertion of ownership, and we know that slavery is wrong. Hence, even if the foetus does have a right to life (and, as we have seen, that assumption is not proven) it does not follow that the foetus has the right to the use and occupancy of the pregnant woman's body. Nor does it follow, as some have claimed,[5] that the foetus is legally entitled to treatments, such as surgery, that the pregnant woman rejects. Although the foetus is not the property of the pregnant woman, neither is the pregnant woman the property of the foetus.

Whenever our culture limits access to abortion on the grounds of alleged foetal rights, it is saying that pregnant women must sacrifice their independence and accept limits on their autonomy and bodily integrity that are required of no other group of citizens, and that are not even required of women themselves after they have given birth.[6] It is this latter point that must be emphasized: foetuses are the only group of entities that have been given legal entitlement to the medical use of the bodies of adult persons. If we are not willing to authorize compulsory 'donations' of blood or organs to save the lives of dying persons, then we should not be willing to tolerate compulsory foetal surgery or Caesarean sections, or to deny abortions.

In discussions of abortion, some defenders of foetal rights suggest that there are no crucial moral differences between the foetus inside the mother's body and the baby born prematurely. It is just a matter of location, they claim, and why should a few inches of travel down the birth canal make any difference to how we treat that being?

In fact, those inches are centrally important when the location is the body of a human person, and when that body is being occupied and used by another entity. No entity has an entitlement to the use of a woman's body, even if that entity is a foetus or embryo at any stage of its development. When extra embryos are produced through *in vitro* fertilization, it would be implausible to claim that they have a right to be reimplanted in the body of the woman from whom the ova were taken. And if a woman candidate for IVF were to die before the implantation of the embryos, no one would argue that the embryos had the right to the occupancy of some other woman's body.[7] Only when the foetus is already occupying a woman's uterus do foetal-rights advocates claim a sort of squatter's rights for it: once the foetus is there, it must be permitted to stay.

There are several reasons why foetal-rights advocates believe that the foetus has squatter's rights to the pregnant woman's body. As a number of feminists have pointed out, discussion of abortion and foetal rights is becoming dominated by metaphors implying that a pregnant woman is like a container or a house inhabited by the foetus. The woman is seen as the 'route to the foetus',[8] or even as a kind of impediment that prevents scientists from seeing the foetus clearly. Ultrasound images and photographic representations of the foetus seem to depict it as independent and self-sufficient; the woman whose body sustains it is nowhere in view.[9] These metaphors reinforce the tendency to see the foetus as having a right to the use of the pregnant woman's body: if that body is just a house or a container,[10] then it is simply a piece of property, with various utilities, that the foetus happens to need for approximately nine months.

In response to these metaphorical depictions of the foetus, it is important to become aware of the ways in which the debates about abortion and alleged foetal rights construct the foetus as a being entirely separate from the pregnant woman. With the development of new forms of reproductive technology, the foetus is seen from a eugenic perspective as an opportunity for actively improving human beings.[11] The possibility of foetal surgery and other forms of intervention seems to transform the foetus into a patient, but a patient with a difference: one to whom access is blocked by the body of the woman, who is often seen as posing a danger to the foetus, whether inadvertently or deliberately.[12]

The foetus is also believed to have interests and needs that are not consistent with the interests and needs of the pregnant woman. Then, because the foetus is small and helpless, it is assumed that it needs an advocate, a more powerful, grown-up human being who will represent its interests and act on its behalf. Some physicians see themselves as the advocates of the foetus; some biological fathers have attempted to assume that role;[13] and those who propose laws to enforce foetal rights and to require foetal treatments would make the state the foetus's advocate.

Once again, some unsupported assumptions are at work here. Why assume that the foetus needs an advocate at all? Such a move is a large and dangerous step towards denying the autonomy of the pregnant woman. To give a physician or spouse or partner the right, as the foetus's representative, to insist on foetal surgery or block an abortion is to hand over control of the woman's body to that physician or spouse. Because pregnancy is an event in a woman's body, the moral relationship of the pregnant woman and the man who impregnated her to the foetus cannot be regarded as symmetrical. Provided the pregnant woman is competent, the responsibility for deciding what happens in and to her body rests with her. This is not to say that she is infallible and cannot make mistakes; it is merely to accord her the same freedom from ownership by others—i.e., slavery—that we accord every member of our culture. Neither the physician nor the spouse has any rights over the woman's body, and that fact should take priority over any claims about the alleged rights of the foetus.

Some people, unsatisfied with these arguments, will be left with the intuition that it is urgent to do something to protect the foetus. That intuition is not entirely unfounded. But I would suggest that we re-evaluate the adversarial and individualistic perspective that pits the foetus against the pregnant woman and makes the physician or sperm provider or state the foetus's advocate. We need to move beyond the isolationist point of view that fails to consider the social context in which women gestate and deliver their babies. We need to understand that protecting and caring for the foetus means protecting and caring for the pregnant woman—through adequate housing, nutrition, education, and medical care, and freedom from physical and emotional abuse. Genuine respect for foetal life would require genuine respect for women. Whenever we reach a stage where interventions in a pregnant woman's body, against her will, allegedly for the sake of the foetus, appear to be necessary, then we should step back and look at the larger picture: in what ways have we failed to support, educate, care for, and appreciate the pregnant woman?

Notes

1See, for example, Sanda Rodgers, 'Fetal Rights and Maternal Rights: Is There a Conflict?' *Canadian Journal of Women and the Law/Revue juridique la femme et le droit* 1 (1986): 456-69; Barbara Katz Rothman, 'Commentary: When a Pregnant Woman Endangers Her Fetus', *Hastings Center Report* 16 (February 1986): 25; and Dawn Johnson, 'A New Threat to Pregnant Women's Autonomy', *Hastings Center Report* 17 (August/September 1987): 33-40.

2'The unborn child is no less a human being whilst residing in his mother's body than he is when he emerges from her body,' according to Morris Schumiatcher, lawyer for Joseph Borowski. Schumiatcher is quoted in Sheilah L. Martin, 'Canada's Abortion Law and the Canadian Charter of Rights and Freedoms', *Canadian Journal of Women and the Law/Revue juridique la femme et le droit* 1 (1986): 348.

[3]See John Woods, *Engineered Death: Abortion, Suicide, Euthanasia and Senecide* (Ottawa: University of Ottawa Press, 1978): 17-61; and Rosalind Hursthouse, *Beginning Lives* (Oxford: Basil Blackwell, 1987): 91-7.

[4]Judith Jarvis Thomson, 'A Defense of Abortion', reprinted in James Rachels, ed., *Moral Problems: A Collection of Philosophical Essays*, second ed. (New York: Harper and Row, 1975), 89-106.

[5]See John A. Robertson and Joseph D. Schulman, 'Pregnancy and Prenatal Harm to Offspring: The Case of Mothers with PKU', *Hastings Center Report* 17 (August/September 1987): 28 and 29; Thomas B. Mackenzie and Theodore C. Nagel, 'Commentary: When a Pregnant Woman Endangers her Foetus', *Hastings Center Report* 16 (February 1986): 24-5; and Julius Landwirth, 'Fetal Abuse and Neglect: An Emerging Controversy', *Pediatrics* 79 (April 1987): 508-14.

[6]Caroline L. Kaufmann, 'Perfect Mothers, Perfect Babies: An Examination of the Ethics of Fetal Treatments', *Reproductive and Genetic Engineering* 1 (1988): 139.

[7]See Christine Overall, *Ethics and Human Reproduction: A Feminist Analysis* (Boston: Allen and Unwin, 1987), 76-9.

[8]Ontario Medical Association Committee on Perinatal Care, 'Ontario Medical Association Discussion Paper on Directions in Health Care Issues Relating to Childbirth' (Toronto, 1984): 12.

[9]Rosalind Pollack Petchesky, 'Foetal Images: The Power of Visual Culture in the Politics of Reproduction', in Michelle Stanworth, ed., *Reproductive Technologies: Gender, Motherhood and Medicine* (Minneapolis: University of Minnesota Press, 1987), 57-80.

[10]See George J. Annas, 'Pregnant Women as Fetal Containers', *Hastings Center Report* 16 (December 1986): 14.

[11]See, for example, Ruth Hubbard, 'Eugenics: New Tools, Old Ideas', in Elaine Hoffman Baruch, Amadeo F. D'Adamo, Jr, and Joni Seager, eds, *Embryos, Ethics, and Women's Rights: Exploring the New Reproductive Technologies* (New York: Haworth Press, 1988), 225-35.

[12]See Robertson and Schulman, 'Pregnancy and Prenatal Harm', 23-33.

[13]See the favourable discussion by Donald DeMarco, in *In My Mother's Womb: The Catholic Church's Defense of Natural Life* (Manassas, VA: Trinity Communications, 1987), 57-60.

Chapter 3

◆ ◆ ◆

Selective Termination in Pregnancy and Women's Reproductive Autonomy

The recent development of a new technological procedure has introduced additional questions to debates about women's reproductive self-determination. Variously called 'selective termination in pregnancy', 'selective reduction of multifoetal pregnancy', or 'selective foetal reduction', the process is performed during the first or second trimester in some instances of multiple pregnancy, either to eliminate a foetus found through prenatal diagnosis to be disabled or at risk of a disability, or simply to reduce the number of foetuses in the uterus. By early 1989, at least two hundred selective termination procedures had been performed around the world.[1]

There are several methods of selective termination, all of which first involve ultrasound imaging to locate the target foetus(es). One method is the transcervical aspiration of amniotic fluid and foetal tissue.[2] Another is the placing of a needle into the foetal thorax until cardiac motion ceases. In the third method, a lethal dose of potassium chloride is injected into the foetal thorax to stop the heart.[3] In the two latter methods the 'terminated' foetus is reabsorbed into the woman's body during the course of pregnancy, and no further surgery is required to remove it from her uterus.

In recent news stories and journal articles some physicians and ethicists have expressed reservations about selective termination, with respect both to its moral justification and to the formation of social policy governing access to and resource allocation for this procedure. For example, Abbyann Lynch, former director of the Westminster Institute for Ethics and Human Values in London, Ontario, has remarked that 'It's like saying to a fetus you are good enough to come on the trip but not make the final voyage',[4] while Margaret Somerville, of the Centre for Medicine, Ethics and Law at McGill University, has stated: 'With abortion, a woman has the right to control over her own body. Selective reduction is different. Control over the body moves to the right to kill a fetus who is competing with another for space. I have a lot of problems with that'.[5]

Some commentators worry that the procedure establishes a precedent for euthanasia. They are also concerned that it will be unjustifiably used by women pregnant with twins who wish to reduce their pregnancy to a singleton, and they therefore recommend restricting availability of the process to multiple pregnancies of three or more.[6] In general, according to Walter Hannah, president of the Canadian Society of Obstetricians and Gynecologists, 'There's no question there should be national guidelines [for selective pregnancy termination]'.[7]

Many discussions of selective termination appear to assume that the procedure is primarily a matter of acting against some foetus(es) on behalf of others. For example, Diana Brahams describes the issue as follows: 'Is it ethical and legally appropriate to carry out a selective reduction of pregnancy—that is, to destroy one or more fetuses in order to give the remaining fetus or fetuses a better chance?'[8] Richard L. Berkowitz et al. pose the problem in the following way: 'Is it justifiable to lower the number of fetuses in the uterus in order to reduce an unspecified risk to all the fetuses?'[9] Similarly, in their report on four selective pregnancy terminations, Mark I. Evans et al. discuss the issue as if the primary choice were the killing or the preservation of the foetuses.[10]

However, this construction of the problem is radically incomplete, for it fails to take into account the women—their bodies and their lives—who should be at the centre of any discussion of selective termination. In fact, selective termination vividly illustrates many of the central ethical and policy concerns that must be raised about the technological manipulation of women's reproductive capacities. When Margaret Somerville expresses concern about 'the right to kill a fetus who is competing with another for space', what she neglects to mention is that the 'space' in question is the pregnant woman's uterus.

According to Evans and colleagues, 'the ethical issues [of selective termination] are the same in multiple pregnancies whether the cause is spontaneous conception or infertility treatment'.[11] Such a claim is typical of many discussions in contemporary bioethics, which abstract specific moral and social problems from the cultural context that produced them. But the issue of selective termination in pregnancy demonstrates the necessity of examining the social and political environment in which issues in biomedical ethics arise.

Selective termination itself must be understood and evaluated with reference to its own particular context. The apparent need or demand for selective termination in fact is created and elaborated in response to prior technological interventions in women's reproductive processes, interventions that are themselves the result of prevailing cultural interpretations of infertility.

Hence it is essential to explore the significance of selective termination for women's reproductive autonomy. The issue acquires added urgency at this point in both Canada and the United States when abortion access and allocation are the focus of renewed controversy. Although not precisely the same as abortion, selective termination is similar in so far as in both cases one or more foetuses are destroyed. The difference is that in abortion pregnancy ends, whereas in selective termination, ideally, it continues, with one or more foetuses still present. I will argue that, provided a permissive abortion policy is justified (that is, a policy that allows abortion at least until the end of the second trimester), a concern for women's reproductive autonomy precludes any general policy restricting access to selective termination in pregnancy, as well as clinical practices that discriminate on non-medical grounds as to which women will be permitted to choose the procedure or how many foetuses they must retain.

♦ ♦ ♦
The 'Demand' for Selective Termination

In recent discussions of selective termination, women with multiple pregnancies are often represented as demanding the procedure—sometimes by threatening to abort the entire pregnancy if they are not allowed selective termination. One television interviewer who talked to me about this issue described women as 'forcing' doctors to provide the procedure. Similarly, a case study of selective pregnancy termination presents a 'Ms Q' who is pregnant with triplets and asks her doctor to 'terminate' two of the foetuses:

> She says she really wants to have a child and 'be a good mother', but doesn't feel capable of caring for more than one child at a time. Even though all three fetuses appear healthy, her preference is to abort all rather than have triplets.[12]

The assumption that individual women 'demand' selective termination in pregnancy places all moral responsibility for the procedure on the women themselves. However, neither the multiple pregnancies nor the 'demands' for selective termination originated out of nowhere. Examination of their sources suggests both that moral responsibility for selective termination cannot rest solely with individual women, and that the 'demand' for selective termination is not just a straightforward exercise of reproductive freedom.

Deliberate societal and medical responses to the perceived problem of female infertility generate much of the need for selective termination, which is but one result of a complex system of values and beliefs concerning fertility and infertility, maternity and children. Infertility is not merely a physical condition; it is both interpreted and evaluated within cultural conditions that help to specify the appropriate beliefs about and responses to the condition of being unable to reproduce. According to the prevailing ideology of pronatalism, women must reproduce, men must acquire offspring, and both parents should be biologically related to their offspring. A climate of acquisition and commodification encourages and reinforces the notion of child as possession.

Infertility is seen as a problem that must be solved by acquiring a child of one's own, biologically related to oneself, at almost any emotional, physical, or economic cost.[13]

The recent increase in numbers of multiple pregnancies comes largely from two steps taken in the treatment of infertility. The use of fertility drugs to prod women's bodies into ovulating and producing more than one ovum at a time results in an incidence of multiple gestation ranging from 16 to 39 per cent.[14] Gamete intrafallopian transfer (GIFT), using several eggs, and *in vitro* fertilization with subsequent implantation of several embryos, to increase the likelihood that the woman will become pregnant, may also result in multiple gestation. 'Pregnancy rate increments are about 8 percent for each pre-embryo replaced in IVF, giving expected pregnancy rates of 8, 16, 24, and 32 percent for 1, 2, 3, and 4 pre-embryos, respectively.'[15]

A 'try anything' mentality is fostered by the fact that prospective IVF patients are often not adequately informed about the very low clinical success rates. In an attempt to improve the chances of success, of the procedure,[16] one physician implants as many as twelve embryos after IVF;[17] a woman who sought selective termination after use of a fertility drug was pregnant with octuplets.[18] Another case reported by Evans and colleagues dramatically illustrates the potential effects of these treatments: one woman's reproductive history includes three Caesarean sections, a tubal ligation, a tuboplasty (attempted repair of the fallopian tubes) after which she remained infertile, IVF with subsequent implantation of four embryos, selective termination of two of the foetuses, revelation via ultrasound that one of the remaining twins had 'severe oligohydramnios [a very small amount of amniotic fluid] and no evidence of a bladder or kidneys', spontaneous miscarriage of the abnormal twin, and intrauterine death of the remaining foetus.[19]

In a commentary critical of selective termination, Angela Holder[20] quotes Oscar Wilde's dictum: 'In this world, there are only two tragedies. One is not getting what one wants, and the other is getting it.' But this begs the question of what it means to say that women 'want' multiple pregnancy, or 'want' selective termination in pregnancy. What factors led these women to take infertility drugs and/or to participate in an IVF program? How do they evaluate fertility, pregnancy, motherhood, children? How do they perceive themselves as women, as potential mothers, as infertile, and where do children fit into these visions? Were they adequately informed of the likelihood that they would gestate more than one foetus? Were they provided with support systems enabling them to understand their own reasons and goals for seeking reproductive interventions, and to provide assistance throughout the emotionally and physically demanding aspects of the treatment? Barbara Katz Rothman's comment regarding women who abort foetuses with genetic defects has more general applicability:

> They are the victims of a social system that fails to take collective responsibility for the needs of its members, and leaves individual women to make impossible choices. We are spared collective responsibility, because we individualize the problem. We make it the woman's own. She 'chooses', and so we owe her nothing.[21]

Uncritical use of the claim that certain women 'demand' selective termination implies that they are just selfish, unable to extend their caring to more than one or two infants, particularly if one has a disability. For example, one physician speaks dismissively of women who are bothered by the 'inconvenience' of a multiple pregnancy.[22] But this interpretation is unjustified. In general, participants in IVF programs are extremely eager for a child. They are encouraged to be self-sacrificing, to acquiesce in the manipulations that the medical system requires their bodies to undergo. As John C. Hobbins notes, these women 'have often already volunteered for innovative treatments and may be desperate to try another'.[23] The little evidence so far available suggests that, by comparison to their male partners, these women are, if anything, somewhat passive in regard to the making of reproductive decisions.[24] There is no evidence that most are not willing to assume the challenges of multiple pregnancy.

An additional cause of multiple pregnancy may be found in conservative attitudes towards the embryo/foetus. One report suggests that multiple pregnancies resulting from IVF are generated not only because clinicians are driven to succeed—and implantation of large numbers of embryos appears to offer that prospect—but also because of 'intimidation of medical practitioners by critics and authorities who insist that all fertilised eggs or pre-embryos be immediately returned to the patient'.[25] Such 'intimidation' does not, of course, excuse clinicians who may sacrifice their patients' well-being. Nevertheless, conservative beliefs in the necessity and inevitability of procreation and the sacredness and 'personhood' of the embryo may contribute to the production of multiple pregnancies.

Thus the technological 'solutions' to some forms of female infertility create an additional problem of female hyper-fertility—to which a further technological 'solution' of selective termination is then offered. Women's so-called 'demand' for selective termination in pregnancy is not a primordial expression of individual need, but a socially constructed response to prior medical interventions.

The debate over access to selective termination exemplifies a classic no-win situation for women, in which medical technology generates a solution to a problem that itself has been generated by medical technology—yet women are regarded as immoral for seeking that solution. While women have been, in part, victimized by reproductive interventions that fail to respect and facilitate their reproductive autonomy, they are nevertheless unjustifiably held entirely responsible for their attempts to cope with the outcomes of these interventions in the forms that are made available to them. From this perspective, selective termination is not so much an extension of women's reproductive choice as it is the extension of control over their reproductive capacity—through the use of fertility drugs, GIFT, and IVF as 'solutions' to infertility that, when successful, often result in multiple gestations; through the provision of a technology, selective termination, to respond to multiple gestation that may create much of the same ambivalence for women as is generated by abortion; and, finally, through the setting of limits on women's access to the procedure.

In decisions about selective termination, women are not simply feckless, selfish, and irresponsible. Nor are they mere victims of their social conditioning and the machinations of the medical and scientific establishments. But they must make their choices in the face of extensive socialization for maternity, a limited range of options, and sometimes inadequate information about outcomes. Women who 'demand' selective termination in pregnancy are attempting to take action in response to a situation not of their own making, in the only way that seems available to them. Hence my argument is not that women are merely helpless victims and therefore must be permitted access to selective termination, but rather that it would be both socially irresponsible and unjust for a health-care system that contributes to the generation of problematic multiple pregnancies to withhold access to a potential, if flawed, response to the situation.

♦ ♦ ♦

A Grim Option

There is reason to believe that women's attitudes towards selective termination may be similar to their attitudes towards abortion. Although abortion is a solution to the problem of unwanted pregnancy, and the general availability of abortion accords women significant and essential reproductive freedom, it is often an occasion for ambivalence, and remains, as Caroline Whitbeck has pointed out, a 'grim option' for most women. It is not something women straightforwardly seek, in the way that they may seek a rewarding career, supportive friends, healthy children, or freer sexuality; rather, it is wanted 'only because of a still greater aversion to the only available alternatives.... [W]omen do not want abortions, although under duress they may resort to them.'[26] Women who abort do, after all, undergo a surgical invasion of their bodies, and some may also experience emotional distress.[27] Moreover, for some women the death of the foetus can be a source of grief, particularly when the pregnancy is wanted and the abortion is sought because of severe foetal disabilities.[28]

Comparable factors may contribute to women's reservations about selective termination in pregnancy. Those who resort to this procedure surely do not desire the invasion of their bodies; nor is it their aim to kill foetuses. In fact, unlike women who request abortions because their pregnancies are unwanted, most of those who seek selective termination are originally pregnant by choice. Such pregnancies are 'not only wanted but achieved at great psychological and economic cost after a lengthy struggle with infertility'.[29]

For such women a procedure that risks the loss of all foetuses, as selective termination does, may be especially troubling. The procedure is still experimental, and its short- and long-term outcomes are largely unknown. According to Evans et al., 'Many more cases will have to be observed to appreciate the true risks of the procedure to both the mother and the remaining fetuses',[30] and Berkowitz et al. state: 'Although the risks associated with selective reduction are known, the dearth of experience with the procedure to date makes it impossible to assess their likelihood.'[31] Evans et al. add: 'any attempt to reduce the number of fetuses [is] experimental and [can] result in miscarriage, and ... infection, bleeding, and other unknown risks [are] possible. If successful, the attempt could theoretically damage the remaining fetuses.'[32]

Note that 'success' in the latter case would be seriously limited, assuming that the pregnant woman's goal is to gestate and subsequently deliver one or more healthy infants. In fact, success in this more plausible sense is fairly low. The success rate for Evans et al. was 50 per cent; for Berkowitz et al., 66.6 per cent. As a consequence, in their study of first-trimester selective terminations, the latter mention the 'psychological difficulty of making the decision [to undergo selective termination]', a difficulty partly resulting from 'emotional bonding' with the foetuses after repeated ultrasound examinations.[33]

Thus women undergoing selective termination, like most of those undergoing abortion, choose a grim option; they are ending the existence of one or more foetuses because the alternatives—aborting all the foetuses (and taking the risk that they will never again succeed in becoming pregnant), or attempting to maintain all the foetuses through pregnancy, delivery, and childrearing—are untenable. Women do not seek selective termination for its own sake, or even simply as a means to an end, but because, as the next section will show, their circumstances and the nature of the pregnancy make any other course of action or inaction unacceptable, morally, medically, or practically.

◆ ◆ ◆
The Challenges of Multiple Gestation

Why don't women who seek selective termination simply continue their pregnancies? John Woods, a philosopher highly critical of abortion, makes the following claim:

Pregnancy does not radically impede locomotion, does not necessarily entail a long-term loss of income, does not disrupt a wide range of social and personal relationships, is not a radical and continuous disturbance, is not a socially anomalous condition, and is not an invasion of privacy.[34]

Every clause in this claim is false. As any mother or pregnant woman could explain, pregnancy and its outcome can and do have all these effects, if not in every case, then in many. Rosalind Hursthouse has remarked: 'Most pregnancies and labours call for courage, fortitude and endurance, though most women make light of them—so why are women not praised and admired for going through them?'[35] No matter how much it is taken for granted, gestating and giving birth to even one child is an extraordinary accomplishment; perhaps even more credit is owed to the woman who bears twins, triplets, or quadruplets. Rather than setting policy limits on women who are not able or willing to gestate more than one or two foetuses, we should be recognizing and understanding the extraordinary challenges posed by multiple pregnancies.

There are good consequentialist reasons why a woman might choose to reduce the number of foetuses she carries. For the pregnant woman, continuation of a multiple pregnancy means 'almost certain preterm delivery, prefaced by early and lengthy hospitalization, higher risks of pregnancy-induced hypertension, polyhydramnios [an over-supply of amniotic fluid], severe anemia, preeclampsia [a potentially dangerous condition involving high blood pressure and water retension], and postpartum blood transfusions'.[36] Another commentator describes the risks for the pregnant woman as including pre-eclampsia, serious postpartum haemorrhage, thrombophlebitis, embolic phenomena, and polyhydramnios.[37] The so-called 'minor discomforts' of pregnancy are increased in a multiple pregnancy, and women may suffer severe nausea and vomiting,[38] or become depressed or anxious.[39] There is also an increased likelihood of Caesarean delivery, entailing more pain and a longer recovery time after the birth.[40]

Infants born of multiple pregnancy risk 'premature delivery, low infant birthweight, birth defects, and problems of infant immaturity, including physical and mental retardation'.[41] There is a high likelihood that these infants 'may be severely impaired or suffer a lengthy, costly process of dying in neonatal intensive care'.[42] Thus a woman carrying more than one foetus also faces the possibility of becoming a mother to infants who will be seriously physically impaired and may not live long.[43]

It is also important to count the social costs of bearing several children simultaneously in a society where the responsibilities, burdens, and lost opportunities occasioned by childrearing fall primarily if not exclusively upon the woman rather than her male partner (if any) or, more equitably, the society as a whole—particularly when the infants are disabled. An article on Canada's first set of 'test-tube quintuplets' reported that the babies' mother, Mae Collier, changed diapers fifty times a day, and went through twelve litres of milk a day and 150 jars of baby food a week. Her husband worked full-time outside the home and 'spen[t] much of his spare time building the family's new house'.[44]

Moreover, while North American culture is strongly pronatalist, it is simultaneously anti-child. One of the most prevalent myths of the West is that North Americans love and spoil their children. In fact, however, a sensitive examination—perhaps from the perspective of a child or a loving parent—of the conditions in which many children grow up gives the lie to this myth.[45] Children are among the most vulnerable victims of poverty and malnutrition. Subjected to physical and sexual abuse, educated in schools that more often aim for custody and confinement than growth and learning, exploited as opportunities for the mass marketing of useless and sometimes dangerous foods and toys, children, the weakest members of our society, are often the least protected. Children are virtually the last social group in North America for whom discrimination and segregation are routinely countenanced. In many residential areas, businesses, restaurants, hotels, and other 'public' places, children are not welcome, and, except in preschools and nurseries, accommodations are rarely made for their physical needs and capacities.

A society that is simultaneously pronatalist and anti-child, and only minimally supportive of mothering, is unlikely to welcome quintuplets and other multiples—except for their novelty—any more than it welcomes single children. The issue, then, is not merely how many foetuses a woman can be required to gestate, but also how many children she can be required to raise, and under what sort of societal conditions.

To this argument it is no adequate rejoinder to say that such women should continue their pregnancies and then surrender some of the infants for adoption by eager childless and infertile couples. It is one thing for a woman to have the choice of making this decision after careful thought and with full support throughout the pregnancy and afterward, when the infants have been given up. Such a choice may be hard enough. It is another matter, however, to advocate a policy that would restrict selective termination in such a way that gestating all the foetuses and surrendering some becomes a woman's only option.

First, the presence of each additional foetus places additional demands on the woman's physical and emotional resources;[46] gestating triplets or quadruplets is not the same as gestating twins. Second, to compel a woman to continue to gestate foetuses she does not want for the sake of other people who do is to treat her as a mere breeder, a biological machine for the production of new human beings.[47] Finally, it would be callous indeed to underestimate the emotional turmoil and pain of the woman who must gestate and deliver a baby only to surrender it to others. In the case of a multiple gestation, additional distress would arise from the necessity of choosing which infant(s) to keep and which to give up.

Hence both physical stress and the social conditions for rearing several infants simultaneously can make untenable the continued gestation of all foetuses in a large multiple pregnancy.

♦ ♦ ♦

Reproductive Rights

Within the existing social context, therefore, access to selective termination must be understood as an essential component of women's reproductive rights. But, as was noted in Chapter 1, in staking reproductive rights claims it is important to distinguish between the right to reproduce and the right not to reproduce. Entitlement to access to selective termination, like entitlement to access to abortion, falls within the right not to reproduce.[48]

Entitlement to choose how many foetuses to gestate, and of what sort, is in this context a limited and negative right. If women are entitled to choose to end their pregnancies altogether, then they are also entitled to choose how many foetuses, and of what sort, they will carry. If it is unjustified to deny a woman access to an abortion of all foetuses in her uterus, then it is also unjustified to deny her access to the termination of some of them. Furthermore, if abortion is legally permitted in cases where the foetus is seriously disabled, it is inconsistent to refuse to permit the termination of one disabled foetus in a multiple pregnancy.

One way of understanding abortion as an exercise of the right not to reproduce is to see it as the premature emptying of the uterus, or the deliberate termination of the foetus's occupancy of it. If a woman has an entitlement to an abortion—that is, to have her uterus emptied of all its occupants—then there is no ground to compel her to maintain all the occupants of her uterus if she chooses to retain only some of them. While the risks of multiple pregnancy for both the foetuses and the pregnant woman increase with the number of foetuses she carries,[49] it does not follow that restrictions on selective termination for pregnancies with smaller numbers of foetuses would be justified. Legal or medical policy cannot consistently tell a woman that she may choose whether to be pregnant, that is, whether her uterus may be occupied, but she may not choose how many foetuses may occupy it.

Foetuses do not acquire a right, either collectively or individually, to use a woman's uterus simply because there are several of them present simultaneously. If one foetus alone in a singleton pregnancy does not have such a right, there is no reason to give several foetuses together, either individually or jointly, such a right.

♦ ♦ ♦
Making Changes

Although I defend a social policy that does not set limits on access to selective termination in pregnancy, there can be no denying that the procedure may raise serious moral problems. For example, as some disabled persons themselves have pointed out, there is a special moral significance to the termination of a foetus with a disability such as Down's syndrome.[50] The use of prenatal diagnosis followed by abortion or selective termination may have eugenic overtones,[51] when the presupposition is that we can ensure that only high-quality babies are born, and that 'defective' foetuses can be eliminated before birth. The foetus is treated as a product for which 'quality control' measures are appropriate. Moreover, since prenatal diagnosis by means of amniocentesis and chorionic villus sampling reveals the sex of offspring, there is also a possibility that selective termination in pregnancy could be used, as abortion already is, to eliminate foetuses of the 'wrong' sex—in most cases, those that are female.[52]

These possibilities are distressing and potentially dangerous to disabled persons of both sexes and to women generally. But the way to deal with these and other moral reservations about selective termination is not to prohibit selective termination or to limit access to it on such grounds as foetal disability or foetal sex choice. Instead, if '[m]any of the currently known instances of grand multiple pregnancies should have never happened'[53] part of the answer is to change the conditions that generate large numbers of embryos and foetuses. The administration of fertility drugs to induce ovulation can be carefully monitored,[54] and in IVF and GIFT procedures more use can be made of the 'natural ovulatory cycle' and of cryopreservation of embryos.[55] The number of eggs implanted through GIFT and the number of embryos implanted after IVF can be limited—not by unilateral decision of the physician, but after careful consultation with the woman about the chances of multiple pregnancy and her attitudes toward it.[56] To that end, there is a need for further research on predicting the likelihood of multiple pregnancy.[57] And, given the experimental nature of selective termination, genuinely informed choice should be mandatory for prospective patients, who need to know both the short- and long-term risks and outcomes of the procedure. Acquiring this information will necessitate the 'long-term follow-up of parents and children ... to assess the psychological and physical effects of fetal reduction'.[58] By these means the numbers of selective terminations can be reduced, and the women who seek selective termination can be protected and empowered.

More generally, however, we should carefully re-evaluate both the pronatalist ideology and the system of treatments for infertility that constitute the context in which selective termination in pregnancy comes to seem essential. There is also a need to improve social support for parenting, and to transform the conditions that make it difficult or impossible to be the mother of triplets, quadruplets, etc., or of a baby with a severe disability. Only through the provision of committed care for children and support for women's self-determination will genuine reproductive freedom and responsibility be attained.

Notes

[1]Dorothy Lipovenko, 'Infertility Technology Forces People to Make Life and Death Choices', *Globe and Mail* (21 Jan. 1989): A4.

[2]Richard L. Berkowitz, Lauren Lynch, Usha Chitkara, et al., 'Selective Reduction of Multifetal Pregnancies in the First Trimester', *New England Journal of Medicine* 118, 16 (1988): 1043-7.

[3]Mark I. Evans, John C. Fletcher, Evan E. Zador, et al., 'Selective First-Trimester Termination in Octuplet and Quadruplet Pregnancies: Clinical and Ethical Issues', *Obstetrics and Gynecology* 71, 3, pt 1 (1988): 289-96.

[4]Lipovenko, 'Infertility Technology', A4.

[5]'Multiple Pregnancies Create Moral Dilemma', *Kingston Whig-Standard* 21 January 1989: 3.

[6]Evans et al., 'Selective First-trimester Termination'.

[7]Lipovenko, 'Infertility Technology', A4.

[8]Diana Brahams, 'Assisted Reproduction and Selective Reduction of Pregnancy', *Lancet* 8572 (1987): 1409.

[9]Berkowitz et al., 'Selective Reduction', 1046.

[10]Evans et al., 'Selective First-trimester Termination'.

[11]Ibid., 293.

[12]Angela R. Holder and Mary Sue Henifin, 'Case Study: Selective Termination of Pregnancy', *Hastings Center Report* 18, 1 (1988): 21.

[13]Christine Overall, *Ethics and Human Reproduction: A Feminist Analysis* (Boston: Allen and Unwin, 1987), Chapter 7.

[14]John C. Hobbins, 'Selective Reduction: A Perinatal Necessity?', *New England Journal of Medicine* 318, 16 (1988): 1062-3.

[15]'Selective Fetal Reduction' (review article), *Lancet* 8614 (1988): 774.

[16]Gena Corea and Susan Ince, 'Report of a Survey of IVF Clinics in the US', in Patricia Spallone and Deborah Lynn Steinberg, eds, *Made to Order: The Myth of Reproductive and Genetic Progress* (Oxford: Pergamon Press, 1987).

[17]Brahams, 'Assisted Reproduction'.

[18]Evans et al., 'Selective First-trimester Termination'.

[19]Ibid., 291.

[20]Holder and Henifin, 'Case Study', 22.

[21]Barbara Katz Rothman, *The Tentative Pregnancy: Prenatal Diagnosis and the Future of Motherhood* (New York: Viking, 1986), 189.

[22]Colum O'Reilly, 'Selective Reduction in Assisted Pregnancies' (letter), *Lancet* 8558: 575.

[23]Hobbins, 'Selective Reduction', 1063.

[24]Judith Lorber, 'In Vitro Fertilization and Gender Politics', in Elaine Hoffman Baruch, Amadeo F. D'Adamo, Jr, and Joni Seager, eds, *Embryos, Ethics, and Women's Rights: Exploring the New Reproductive Technologies* (New York: Haworth Press, 1988).

[25]'Selective Fetal Reduction', 774.

[26]Caroline Whitbeck, 'The Moral Implications of Regarding Women as People: New Perspectives on Pregnancy and Personhood', in William B. Bondeson, H. Tristram Engelhardt, Jr, Stuart F. Spicker, and Daniel H. Winship, eds, *Abortion and the Status of the Fetus* (Boston: Reidel, 1984), 251-2.

[27]Kathleen McDonnell, *Not an Easy Choice: A Feminist Re-examines Abortion* (Toronto: Women's Press, 1984), 33-6.

[28]Rothman, 'Tentative Pregnancy', Chapter 7.

[29]Evans et al., 'Selective First-trimester Termination', 292.

[30]Ibid.

[31]Berkowitz et al., 'Selective Reduction', 1046.

[32]Evans et al., 'Selective First-trimester Termination', 290.

[33]Berkowitz et al., 'Selective Reduction', 1046.

[34]John Woods, *Engineered Death: Abortion, Suicide, Euthanasia and Senecide* (Ottawa: University of Ottawa Press, 1978), 80.

[35]Rosalind Hursthouse, *Beginning Lives* (Oxford: Basil Blackwell, 1987), 300.

[36]Evans et al. 'Selective First-trimester Termination', 292.

[37]Hobbins, 'Selective Reduction'.

[38]Alastair H. MacLennan, 'Multiple Gestation: Clinical Characteristics

and Management', in Robert K. Creasy and Robert Resnick, eds, *Maternal-Fetal Medicine: Principles and Practices*, second ed. (Philadelphia: W.B. Saunders, 1989).

[39]Jose C. Scerbo, Powan Rattan, and Joan E. Drukker, 'Twins and Other Multiple Gestations', in Robert A. Knuppel and Joan E. Drukker, eds, *High-risk Pregnancy: A Team Approach* (Philadelphia: W.B. Saunders, 1986).

[40]MacLennan, 'Multiple Gestation'.

[41]'Selective Fetal Reduction', 773.

[42]Evans et al., 'Selective First-trimester Termination', 295.

[43]MacLennan, 'Multiple Gestation'.

[44]Victoria Stevens, 'Test-tube Quints Celebrate First Birthday', *Toronto Star* (6 Feb. 1989): A7.

[45]Letty Cottin Pogrebin, *Family Politics: Love and Power on an Intimate Frontier* (New York: McGraw-Hill, 1983).

[46]MacLennan, 'Multiple Gestation'.

[47]Gena Corea, *The Mother Machine: Reproductive Technologies from Artificial Insemination to Artificial Wombs* (New York: Harper and Row, 1985); Margaret Atwood, *The Handmaid's Tale* (Toronto: McClelland and Stewart, 1985).

[48]Overall, *Ethics and Human Reproduction*.

[49]MacLennan, 'Multiple Gestation'.

[50]Deborah Kaplan, 'Disability Rights Perspectives on Reproductive Technologies and Public Policy', in Sherrill Cohen and Nadine Taub, eds, *Reproductive Laws for the 1990s* (Clifton, NJ: Humana Press, 1989); Adrienne Asch, 'Reproductive Technology and Disability', in ibid.; Marsha Saxton, 'Prenatal Screening and Discriminatory Attitudes about Disability', in Baruch et al., eds, *Embryos, Ethics, and Women's Rights*.

[51]Ruth Hubbard, 'Eugenics: New Tools, Old Ideas', in Baruch et al. eds, *Embryos, Ethics, and Women's Rights*.

[52]Helen B. Holmes and Betty B. Hoskins, 'Prenatal and Preconception Sex Choice Technologies: A Path to Femicide?' in Gena Corea et al., eds, *Man-made Women: How New Reproductive Technologies Affect Women* (Bloomington: Indiana University Press, 1987); Madhu Kishwar, 'The Continuing Deficit of Women in India and the Impact of Amniocentesis', in ibid.; Roberta Steinbacher and Helen B. Holmes, 'Sex Choice: Survival and Sisterhood', in ibid.; Robyn Rowland, 'Motherhood, Patriarchal Power, Alienation and the Issue of "Choice" in Sex Preselection', in ibid.

[53]Evans et al., 'Selective First-trimester Termination', 296.

[54]Hobbins, 'Selective Reduction'.

[55]'Selective Fetal Reduction', 774.

[56]Brahams, 'Assisted Reproduction'.

[57]Ian Craft, Peter Brinsden, Paul Lewis, et al., 'Multiple Pregnancy, Selective Reduction, and Flexible Treatment' (letter), *Lancet* 8619 (1988): 1087.

[58]'Selective Fetal Reduction', 775.

Chapter 4

◆ ◆ ◆

Biological Mothers
and the Disposition of Foetuses
after Abortion

This chapter begins from abortion, but is not about abortion. Rather, it examines some aspects of the role of the biological mother with respect to the disposition of foetuses after abortions. Should the biological mother be entitled to determine the disposition of the foetus, where that disposition is not in the interests of the foetus itself? In particular, should the mother be entitled to be assured of the foetus's death, either during or after the abortion? Is it wrong to preserve, against the woman's wishes, foetuses that do or can survive abortion?

While the numbers of foetuses that survive abortion are small, the question of their disposition is significant because it brings together some central problems concerning maternal/foetal relationships, reproductive autonomy and control over the body, and the moral status of the foetus. Before pursuing these, it is necessary to consider the nature of the question.

First, it might seem inappropriate to use the term 'foetus' with respect to an entity that survives abortion. In legal and medical contexts, 'foetus' is generally used to refer only to the 'product of conception' *before* it has emerged from its mother's body.[1] As a result, the terms 'abortus' and 'infant' are sometimes used to describe the entity that survives abortion. But their use appears to beg some of the moral questions at issue: 'abortus' implies that the entity is mere aborted tissue, whereas 'infant' suggests that it is no different from any other baby. While one or the other of these two views may be correct, it is preferable not to prejudge the issues, and therefore to use the more neutral term 'foetus'.[2]

Second, it might be asked whether a foetus can have interests, especially interests independent of those of the pregnant woman. Indeed, given the uniquely close connection between pregnant woman and foetus, it may not be appropriate to distinguish between foetal and maternal interests during the pregnancy itself, or at least to regard them as being in conflict. But when pregnancy ends and the foetus is no longer *in utero*, it is entirely reasonable, as later discussion will confirm, to speak of the foetus as having interests independent of those of the woman. In this discussion it is assumed that the foetus does have interests, after its emergence from the uterus, at least by virtue of its sentience or its potential for sentience.[3]

Third, the issue here primarily concerns relatively developed foetuses of about five months' or more gestational age. Both because of their extreme gestational immaturity, and because of the destructive nature of most abortion procedures, very few such foetuses actually do survive abortion, at present; that is, very few continue to show a sustained heartbeat, spontaneous respiration, and muscle movement. However, if and when technological advancements permit the time of viability to be moved downward, then moral questions about entitlements of disposition will also arise for younger foetuses that survive abortion.

Viability itself is not the critical moral criterion for determining the disposition of the foetus, since viability is a technologically-dependent criterion, having little or nothing to do with the foetus itself, the pregnant woman, or the maternal/foetal relationship. Nor is viability a morally significant point at which abortions should be forbidden.[4] If the aim is foetal preservation, there is a moral paradox in permitting abortions before but not after viability. By definition, abortion before viability causes the death of a foetus that is incapable of survival outside the uterus. Once a foetus is capable of independent survival, the preservation of its existence is not incompatible with its removal from the uterus. Hence preventing abortions after viability does not serve the goal of foetal protection—although it does, inevitably, violate the reproductive autonomy of those who need late abortions.

Fourth, debates about the alleged personhood of foetuses have so far seemed to be of little help in the controversies over abortion. The discussion that follows assumes both that the foetus is not yet a person, although it has the capacity to become one, and that a being need not be a person in order to have interests. It also assumes that a pro-choice position with respect to abortion is morally justified.

Finally, the investigation of these questions has further implications with respect to the potential use of foetuses and foetal tissue for research or therapeutic purposes. However, the focus in this chapter will be on the preservation of foetuses for the sake of an ongoing life of their own, rather than for use as research material.

♦ ♦ ♦
The Problem

It is possible to distinguish between two different concepts of abortion: abortion as the termination of the pregnancy, and abortion as the killing of the foetus.[5] While the two are ordinarily empirically linked, they are conceptually and morally distinct. By emphasizing the first concept, one could interpret a decision to have an abortion primarily as the choice to have the foetus removed from the uterus and to surrender it, somewhat in the way that children are surrendered in adoption.[6] On this view, the woman who aborts is entitled to have the foetus removed from her body, but not to have it dead. She may want it dead, but whether in general women who abort want this is not clear; certainly it is not always the case:

> In some cases, abortions are induced because the continuation of the pregnancy poses a serious threat to the woman's life or health. What is intended then is to restore the woman's health or save her life by interrupting the pregnancy, severing the tie between the woman and her fetus. If the tie could be severed without

terminating fetal life, this would be the preferred outcome. Fetal death may thus be a foreseen but unintended consequence of an abortion that is therapeutic for the pregnant woman.[7]

Even if the woman does want the death of the foetus, Sissela Bok and others have argued that this is not a desire that should be gratified:

> [W]hile a woman does have the right to an abortion in the sense of the termination of her pregnancy, she does not have the right to the death of the fetus. The termination of early pregnancy carries with it, at present, fetal failure to survive. But in later pregnancy, where abortion and death of the fetus do not necessarily go together, it is a fallacy to believe that a right to the first also implies a right to the second.[8]

Defenders of this view of abortion as removal of the foetus must acknowledge the practical and moral difficulties that attempting to save aborted foetuses could raise, especially with regard to the possible survival of injured and damaged foetuses. The difficulties of 'salvaging' very immature foetuses, the subsequent pain caused to them, and possible resulting disabilities, should not be underestimated.

By contrast, other philosophers have emphasized the second concept of abortion, the killing of the foetus, suggesting that the death of the foetus is an inherent component of full abortion rights. 'The quest for abortion rights for women is not merely a quest for control of one's body, though it is surely this in part. The quest for abortion rights is also a quest for the right to terminate the development of an unwanted foetus[,] which can not be accomplished without killing it'.[9] Hence the questions posed at the beginning of this chapter: Should the pregnant woman be entitled to be assured of the foetus's death, either during or after the abortion? Is it wrong to preserve, against the woman's wishes, foetuses that do or can survive abortion?

In Chapter 2 I suggested that, since certain social practices imply that the foetus is not a person with a right to life, the burden of proof rests on those who claim that it is a person. Nonetheless, depriving a being of life (even when that being is not a person) appears *prima facie* to be wrong and to call for justification. Hence, while abortion is readily justified on the grounds of respect for women's reproductive autonomy, the burden of proof here rests upon those who claim, in addition, an entitlement to the foetus's death in those cases where it survives the abortion procedure. The remainder of this chapter presents and evaluates some major arguments, primarily but not exclusively advanced from a feminist perspective, suggesting that failure to observe the wishes of the biological mother with respect to the death of the foetus is morally wrong.

◆ ◆ ◆

Argument I

To keep the foetus alive against the wishes of its mother—in the social rather than the biological sense of that term—is a violation of the woman's reproductive autonomy. A woman who aborts wants not only not to be pregnant but also not to be the mother of this particular foetus, and hence to the child it could become.[10] For reasons having to do with her economic situation, her health, her psychological state, her job, her education, her relationships, or her commitments to other children, the woman wants to prevent the possible existence of the child that this foetus could become. The entitlement to abort is the entitlement to decide whether or not to be a mother to the child that this foetus could become.[11]

Respect for the woman's reproductive autonomy therefore precludes both 'saving' the foetus against the woman's wishes and the (for now) science-fiction scenario of transplantation of the foetus to another woman's uterus or to an artificial uterus (so-called 'ectogenesis'). As Steven Ross points out, if women going to abortion clinics were told that their foetuses would be removed from their bodies, without harm to themselves or to the foetuses, and kept alive elsewhere for the rest of gestation, many would not be satisfied:

> What they want is not to be saved from the 'inconvenience of pregnancy' or 'the task of raising a certain (existing) child'; what they want is *not to be parents*, that is they do not want there to *be*

a child they fail or succeed in raising. Far from this being 'exactly like' abandonment, they abort precisely to avoid being among those who later abandon. They cannot be satisfied *unless* the fetus is killed; nothing else will do.[12]

Moreover, under present circumstances, the attempt to save the foetus born alive after an abortion 'does not prevent a woman from having an abortion but threatens her with a brain-damaged infant'.[13] In such a case, the woman who seeks to end her reproductive activity through abortion is forced to become the mother of an infant that may well be badly damaged by the abortion procedure itself.

Response

In having the abortion, the woman has surrendered the foetus; she has chosen not to become a social mother to it. Hence, if the foetus is saved, she ought not to be compelled to become the social mother of the foetus, whatever its medical condition turns out to be. It is no longer her responsibility to rescue, preserve, or nurture it; she no longer has any ongoing special moral commitment to it, just by virtue of being its biological mother. If this is acknowledged, then the woman's reproductive autonomy—in particular, whether or not she will become a mother in the social sense—would not be violated by saving the foetus. She would remain only its genetic mother, a connection that is unavoidable once conception has occurred.

What is not clear, however, is whether this response unjustifiably trivializes genetic parenthood. The point of Argument 1 is that if the foetus is saved, there will now exist a human being genetically related to the woman.[14] In an important way, this human being is no mere stranger. As Ross argues, this child could represent a kind of failure for the woman, a failure to 'be a certain kind of *person*, that is, the sort who has children only when able to raise them oneself in an environment one finds right'.[15] Such a woman may feel and believe that she should be the one to raise any children she has borne; therefore, if the foetus is preserved, '[t]here would always be in the world a person to whom one was failing to be a proper or full parent, and this is a failure one understandably dreads.' In addition, saving the foetus would mean subjecting the biological mother to a form of compulsory adoption, with all the potential for suffering that is experienced by women who give up offspring. 'Although we [the pregnant woman] would not be bringing the child up, because someone else (let us assume) is all too gladly embracing those tasks, we do not want precisely this state of affairs to come about'.[16]

♦ ♦ ♦
Argument II

Saving the foetus against the mother's will would be like compelling her to donate organs, blood, or gametes against her will. If the compulsory 'donation'—'procurement' is probably a more appropriate word[17]—of bodily parts and products such as organs, blood, or gametes, is neither morally justified nor legally permitted, then foetuses, which are equally body products, ought not to be taken from women against their will.[18] To save the foetus against the mother's will violates a fundamental principle of medical ethics: informed consent.

This argument becomes especially urgent if we imagine that foetal viability is pushed back, through technological advancements in neonatal intensive care, earlier and earlier in the process of gestation. Such foetuses could then be 'rescued' after abortion, even when there is no evidence for their sentience. Indeed, the case for such 'rescue' has already been advanced by anti-abortionists:

> Work has already been done toward the development of artificial placentas, and [Bernard] Nathanson sees the possibilities for the rescue of embryos prefigured in the remarkable advent of fiber optics and microsurgery.... What remains, then, as far as the embryo is concerned, is the development of 'an instrument of sufficient delicacy that it can be threaded through the hysteroscope ... and can then pluck [the new being] off the wall of the uterus like a helicopter rescuing a stranded mountain climber.[19]

Hadley Arkes predicts that 'the law' could compel a woman to have her embryo removed in the first few weeks of pregnancy, once technology reaches the point where it is 'possible to rescue the child'.[20]

This language of 'mountain climbers' and 'rescue' suggests, falsely, that a small but sentient and threatened *person* cowers in the woman's uterus, awaiting its salvation by benign medical technology. It is no coincidence that anti-abortionists present their own efforts to prevent abortions as the 'rescue' of babies.[21] From the point of view taken here, that the foetus is not a person, Arkes's proposal sounds like assault upon and invasion of a woman's body, and theft of a component of it. Women would effectively be coerced into being 'foetus farms', and their bodily control would be severely compromised.

Response

Just as people ought not to be compelled to undergo surgery against their will, so also abortions ought never to be compulsory. Women's right to refuse medical interventions on their own bodies must be respected. But providing that the abortion itself is freely chosen, not compelled, it is in this respect not analogous to compulsory organ 'donation' or procurement.

Further, there are limits to the organ analogy. Foetuses are in women's bodies, but not part of their bodies; they are arguably not a renewable resource in the same way that blood and sperm are. Unlike bodily organs, they can, later in their development, survive independently of the woman's body and become persons, and, at least late in gestation, they are sentient. 'The fetus is a developing being and potential member of the human community'; it has 'a unique genetic identity, a species-specific physical appearance, and a truncated participation in human social relations'.[22] Pregnant women do not seem to experience the foetus as just another bodily organ; they often experience it as both part of and different from themselves, and they sometimes develop a type of relationship with the foetus before its birth. By contrast, human beings do not usually develop relationships with their bodily organs. Moreover, while organs serve some purpose in the person's body, so that if they are removed for non-therapeutic reasons there is a deficit in the body, this is not the case for the foetus. The woman needs her bodily organs, but she does not need the foetus: the foetus needs her. These characteristics of foetuses, as opposed to bodily organs, provide all the more reason to be concerned about the fate of the foetus that survives abortion, and they discredit the analogy between foetuses and bodily organs.

In addition, the analogy to compulsory organ procurement appears, unjustifiably, to imply that the woman owns the foetus: 'In this paradoxical morality there [is] a curious assertion of "property rights": it was somehow easier to kill the fetus in the womb than to give away to others what was recognizably a child—and recognizably, also, a child of one's "own".'[23]

Interestingly, some commentators have been willing to see foetuses, along with organs, limbs and bodily fluids, as property[24] over which some limitations of disposal may apply. But even if organs, limbs, and fluids should be seen as property (and there may be good arguments against it), the disanalogies between foetuses and bodily parts suggest that foetuses cannot likewise be regarded as property.

The purview of informed consent is justifiably limited with respect to blood and gametes, and perhaps it ought likewise to be limited with respect to foetuses. We are not entitled to say who ought or ought not to receive our blood or gametes. We are not allowed to make invidious exclusions with respect to their use: for example, racists are not entitled to specify that only whites may have their blood; the homophobic are not entitled to specify that only heterosexuals may have their gametes (although this latter principle often governs the practices of sperm banks). A comparable limitation on the range of informed consent with respect to the disposition of foetuses may therefore seem morally justified.

Nevertheless, there is a significant moral and social policy question concerning whether we should in fact have more control over the disposition of bodily parts and products. In a recent, much disputed case, a man's cancerous blood and tissue were surgically removed and subsequently cultured and developed, allegedly without his knowledge or explicit consent, to develop a patented and commercially valuable cell line.[25] The judicial assessment of the case manifested 'the irony of the conclusion that everyone *except* the patient can own the patient's removed cells and treat them as property'.[26] Such disputes at least indicate that it may be important to grant individuals more knowledge and decision-making about and control over the disposition of material removed from their bodies, even if that material merely seems to the patient to be 'waste'.

The case against the saving of foetuses that survive abortion need not rest upon seeing the foetus as property in the same way that jewellery or other possessions are property. To say that 'X is mine' is sometimes to make an ownership claim. Sometimes, however, it is to claim X as my responsibility, or as subject to my decision-making. For example, when a woman describes offspring as 'my children', she is not claiming that she owns them; rather she is asserting responsibility for and connection to them. Thus Susan Sherwin argues that

> women are in a privileged position with respect to the fetuses developing in their bodies, and ... in most circumstances, they are entitled to decide the future of those fetuses. This is not because they own the fetuses, for they ought not to be free to sell them, but because they are responsible for them and should be trusted to decide if continued life when removed from the womb is in the best interest of the fetus.[27]

Rejecting the compulsory preservation of foetuses, therefore, need not assume a property relationship between the pregnant woman and the foetus. When a woman chooses abortion, she does not necessarily choose to have her foetus 'snatched' from her; if she does not, then preservation of the foetus would be comparable to a compulsory organ 'donation' in which the patient chooses organ removal but does not agree to the subsequent salvaging and use of the organ.[28]

♦ ♦ ♦
Argument III

By virtue of her physical relationship to it, the biological mother is the most appropriate person—perhaps the only one—to decide the disposition of the foetus. As Mary Anne Coffey puts it: 'If her child is dying of a fatal illness ... a mother now can direct that the child not be resuscitated. Why deny a woman the same right when a fetus that survives an abortion dies?'.[29] Susan Sherwin adds that it is the woman's 'vision of threats facing the developing child' that might motivate her to want the foetus destroyed:

> An analysis attentive to the interests of children and women ... must recognize that protecting the interests of the embryo does not necessarily mean preserving its life.... [I]t is legitimate ... that the person who has the most intimate relationship with the fetus and who has the most invested in its development—i.e., the mother—should be the one to decide on how its interests may best be served.[30]

Response

It must be acknowledged that the pregnant woman is the best and only person to make decisions about the foetus while it is *in utero*. But to the extent that a foetus *ex utero* is comparable to a premature baby, as Argument III appears to grant, the interests of the offspring should be the prevailing criterion for decision-making about it. It is improbable to suppose that a foetus that survives abortion is always better off dead (though it may often be);[31] moreover, there may be adoptive parents willing to raise it. Hence giving the biological mother the entitlement to the death of the foetus would sometimes mean overlooking the foetus's interests in a way that would not be condoned for premature infants; such a choice would not be morally justified.

Should the foetus that survives abortion be regarded as an ordinary newborn, or rather, as a premature newborn? In terms of its physical characteristics it may be very like such a newborn, provided it has not suffered injuries in the process of the abortion. Its 'arrival' into the world is induced rather than the outcome of the natural course of labour—but many other births are induced through the use of pitocin drips or Caesarean sections. Moreover, like an ordinary newborn, the foetus that survives abortion 'may have no intrinsic properties that can ground a moral right to life stronger than that of a fetus just before birth, [but] its emergence into the social world makes it appropriate to treat it as if it had such a stronger right'.[32] The point here is *not* that the fact of birth (or removal from the uterus during abortion) constitutes the foetus as a person, but that it does become a biologically separate human being. As such, it can be known and cared for as a particular individual. It can also be vigorously protected without negating the basic rights of women. There are circumstances in which infanticide may be the best of a bad set of options. But our own society has both the ability and the desire to protect infants, and there is no reason why we should not do so.[33]

◆ ◆ ◆

Argument IV

Deliberately withholding the determination of the disposition of the foetus from the biological mother is yet another example of the takeover of reproduction from women. 'The extent to which the rights of women are diminished in abortion policy and litigation, when the fetus is part of the woman's body, should make us seriously question the extent to which they will be further diminished as the fetus is removed from the female body'.[34] Susan Sherwin argues that without the freedom to decide the fate of their foetuses, 'women will not have the reproductive freedom necessary, and, in particular, they will certainly have difficulty in getting abortions'.[35] Anne Donchin argues that the technological maintenance, outside the woman's body, of foetuses that survive abortion would be a manifestation of distrust of women's bodies; laboratory technicians are not likely to do 'a more competent job of gestation than pregnant women'. Moreover, she asks,

> if extrauterine gestation were to become an established practice, would not many women be pressured to adopt it— 'for the good of their baby'?
>
> Though abortion may count as a harm to the fetus, laboratory gestation would as well—not only to particular 'unwanted' fetuses but to *all* future fetuses. For, within the prevailing social framework, once the practice was established it is unlikely that only intentionally aborted fetuses would be nourished in laboratories. Any other fetus considered 'at risk' for any reason would count as a potential beneficiary of laboratory observation and intervention.[36]

Response

It must be granted that the appropriation of reproductive control from women must be resisted. To this end, women are entitled to have the type of abortion they choose (within the limits of good medical practice). This means that they are entitled to choose, if they wish, means of abortion that will likely produce the death of the foetus; they are not required to choose abortifacients that will preserve its life. Pregnant women are not morally required to exhibit 'moral heroism' by putting their own lives at risk for the sake of a possibly viable foetus.[37] Nor are women compelled to undergo less safe forms of abortion in order to provide intact foetal tissue for purposes of transplant or research. The availability of these choices and protections with regard to the process or means of abortion respects women's bodily autonomy. In addition, the encroachment of the state and the medical profession on women's reproductive autonomy during pregnancy must be adamantly resisted,[38] along with legal attempts to compel prenatal treatment, to administer forced Caesareans, or 'take custody of the fetus' before birth.[39]

It is therefore necessary to reject the views of philosophers such as David S. Levin, who claims that if there is the possibility of keeping a foetus alive after it is removed from a woman's body, then she has a 'minimal responsibility' of allowing the being to be removed alive.[40] Levin states that 'if and when removal without killing becomes possible, her [the pregnant woman's] right to control her own body cannot justify killing the foetus'.[41] But without any specification of time or place, the phrase 'killing the fetus' is ambiguous. If it means killing the foetus *in utero*, as a consequence of a particular abortion operation, then the killing is justifiable by reference to the woman's control over her own body. Only if the killing of the foetus takes place after it has been removed from the woman's body is it no longer justified simply by reference to the woman's control over her own body, or to the fact that she is the biological mother of the foetus.

◆ ◆ ◆
Conclusion

This examination of the alleged rights of biological mothers to determine the disposition of the foetus after abortion suggests that it is essential to make a distinction between two different questions. First, who should decide about the disposition of the foetus? And second, does the pregnant woman who seeks an abortion have an entitlement to the death of the foetus?

As the discussion here has suggested, the pregnant woman should decide the disposition of the foetus. For the pregnant woman is, in Sherwin's words, 'the person who has the most intimate relationship with the fetus and who has the most invested in its development'.[42] Given the history of the appropriation of women's reproductive autonomy by male partners and by members of the medical establishment, it is deeply problematic to assign this responsibility to the biological father or to physicians. As John Robertson suggests, '[I]n cases of conflict between her [the pregnant woman] and the father over disposition, one could argue that her interests control because the fetus was removed from her body.'[43]

But while no person has a greater entitlement than the mother to decide about the fate of the foetus that survives abortion, and the decision therefore belongs to her, it does not follow that the woman's decision is necessarily correct. In particular, the choice of death for the foetus is not rendered morally correct simply because the decision is made by its biological mother. Though the pregnant woman is entitled to forms of abortion that may result in the death of the foetus *in utero,* she does not have an *entitlement* to the death of the foetus if it survives abortion. The arguments canvassed in this chapter do not establish the truth of the claim that the woman is automatically entitled, by virtue of being the biological mother, to have the foetus die after it is removed from her uterus, or that her wishes about its disposition, where that disposition is not in the foetus's interests, are necessarily morally justified.

What, then, should be done with foetuses that survive abortion? The answer is that their mothers' decisions on their behalf should be guided by the interests of the foetuses, just as they would be for other premature infants. This is not to say that foetuses must inevitably be preserved and protected, or that extraordinary medical measures must necessarily be taken on their behalf, for sometimes, perhaps often, the foetus that survives abortion is better off dead. Nor is it to say that women should be compelled to raise the foetuses they have aborted, for a decision to abort is (in part) a considered choice not to be the mother of the child this foetus could become. Nor, finally, is it to say that the decision about the disposition of foetuses must devolve upon the male progenitor or the physician(s), for there is no case for their entitlement that overrides that of the pregnant woman. It is to say, however, that the burden of proof still rests on those individuals, including biological mothers, who wish to kill the foetus, to let it die, or to treat it in ways that are not in its interests.

Notes

1. Law Reform Commission of Canada, *Crimes Against the Foetus*, Working Paper 58 (Ottawa, 1989).

2. Moreover, there is a growing body of bioethical literature that uses the term 'foetus' to refer to the entity that survives abortion.

3. On the significance of sentience see Mary Anne Warren, 'The Moral Significance of Birth', *Hypatia* 4, 3 (Fall 1989): 49-52.

4. Christine Overall, *Ethics and Human Reproduction: A Feminist Analysis* (Boston: Allen and Unwin, 1987), Chapter 4.

5. Overall, *Ethics*.

6. Raymond M. Herbenick, 'Remarks on Abortion, Abandonment, and Adoption Opportunities', *Philosophy and Public Affairs* 5, 1 (1975): 98-104; this analogy may not be completely appropriate in cases of abortion for foetal abnormality. In such cases, the foetus is very much wanted. Yet the foetus itself may be better off dead, not preserved. On the ambiguities of seeking abortion for the benefit of the foetus, see Paul F. Camenisch, 'Abortion: For the Fetus's Own Sake?' in John E. Thomas, ed., *Medical Ethics and Human Life* (Toronto: Samuel Stevens, 1983), 135-43.

7. Mary B. Mahowald, Robert A. Ratcheson, and Jerry Silver, 'The Ethical Options in Transplanting Fetal Tissue', *Hastings Center Report 17, 1* (February 1987): 13.

8. Sissela Bok, 'The Unwanted Child: Caring for the Fetus Born Alive

After an Abortion', in Carol Levine and Robert M. Veatch, eds, *Cases in Bioethics*, rev. ed. (Hastings-on-Hudson, NY: Hastings Center, 1984), 2.

[9]David S. Levin, 'Thomson and the Current State of the Abortion Controversy', *Journal of Applied Ethics* 2, 1 (1985): 125; cf. Steven L. Ross, 'Abortion and the Death of the Fetus', *Philosophy and Public Affairs* 11, 3 (1982): 236, and Herbenick, 'Remarks', 101.

[10]I owe this argument to Lois Pineau.

[11]Daniel I. Wikler, 'Ought We to Try to Save Aborted Fetuses?' *Ethics* 90 (October 1979): 58-65.

[12]Ross, 'Abortion and the Death of the Fetus', 238, his emphasis.

[13]David C. Nathan, 'The Unwanted Child: Caring for the Fetus Born Alive after an Abortion', in Levine and Veatch, eds, *Cases in Bioethics*, 4.

[14]For further discussion of the interest in avoiding genetic offspring, see John A. Robertson, 'Resolving Disputes over Frozen Embryos', *Hastings Center Report* 19, 6 (November/December 1989): 7-12.

[15]Ross, 'Abortion and the Death of the Fetus', 241, his emphasis.

[16]Ibid., 239.

[17]Janice C. Raymond, 'Reproductive Gifts and Gift Giving: The Altruistic Woman', *Hastings Center Report* 20, 6 (November/December 1990): 7-11.

[18]A version of this argument was presented to me by Sanda Rodgers.

[19]Hadley Arkes, *First Things: An Inquiry into the First Principles of Morals and Justice* (Princeton: Princeton University Press, 1986), 377.

[20]Ibid., 378.

[21]Gary Leber, 'We Must Rescue Them', *Hastings Center Report* 19, 6 (November/December 1989): 26-7.

[22]Kathleen Nolan, 'Genug Ist Genug: A Fetus Is Not a Kidney', *Hastings Center Report* 18, 6 (December 1988): 16.

[23]Arkes, *First Things*, 371.

[24]Lori B. Andrews, 'My Body, My Property', *Hastings Center Report* 16, 5 (October 1986): 28-38.

[25]Ibid.; George J. Annas, 'Whose Waste Is It Anyway? The Case of John Moore', *Hastings Center Report* 18, 5 (October/November 1988): 37-9; and Annas, 'Outrageous Fortune: Selling Other People's Cells', *Hastings Center Report* 20, 6 (November/December 1990): 36-9.

[26]Annas, 'Outrageous Fortune', 37, his emphasis.

[27]Susan Sherwin, review of Overall, *Ethics and Human Reproduction*, in *Atlantis* 13, 2 (Spring 1988): 125.

[28]There are further feminist reasons for avoiding the ownership paradigm

for the foetus, since there is a developing history of seeing the foetus as the property of the male progenitor, the man's 'baby'.

[29]Mary Anne Coffey, review of Overall, *Ethics and Human Reproduction*, in *Resources for Feminist Research/Documentation sur la recherche féministe* 18, 1 (March 1989): 11.

[30]Sherwin, review, 125.

[31]Ibid.

[32]Warren, 'The Moral Significance of Birth', 57.

[33]Ibid., 62.

[34]Janice C. Raymond, 'Of Ice and Men: The Big Chill over Women's Reproductive Rights', *Issues in Reproductive and Genetic Engineering: Journal Of International Feminist Analysis* 3, 1 (1990): 49.

[35]Sherwin, review, 125.

[36]Anne Donchin, 'The Growing Feminist Debate over the New Reproductive Technologies', *Hypatia* 4, 3 (Fall 1989), 144.

[37]LeRoy Walters, 'The Unwanted Child: Caring for the Fetus Born Alive After an Abortion', in Levine and Veatch, eds, *Cases in Bioethics*, 6.

[38]Janet Gallagher, 'Fetus as Patient', in Sherrill Cohen and Nadine Taub, eds, *Reproductive Law for the 1990s* (Clifton, NJ: Humana Press, 1989), 185-235; National Association of Women and the Law, 'A Response to *Crime Against the Foetus*, The Law Reform Commission of Canada's Working Paper #58' (Ottawa, 1989).

[39]Maggie Thompson, 'Whose Womb Is It Anyway?' *Healthsharing* (Spring 1988): 14-17.

[40]Levin, 'Thompson', 124.

[41]Ibid., 125; cf. Ellen Frankel Paul and Jeffrey Paul, 'Self-Ownership, Abortion and Infanticide', *Journal of Medical Ethics* 5 (1979): 135.

[42]Sherwin, review, 125.

[43]John A. Robertson, 'Rights, Symbolism, and Public Policy in Fetal Tissue Transplants', *Hastings Center Report* 18, 6 (December 1988): 9.

Chapter 5

♦ ♦ ♦

Frozen Embryos and 'Fathers' Rights': Parenthood and Decision-making in the Cryopreservation of Embryos

This chapter will examine some aspects of the current practice of cryopreservation of 'spare' human embryos that are 'left over' after *in vitro* fertilization.[1] This practice raises a number of significant issues in applied ethics, many of them relating to the general justifiability of cryopreservation. On the one hand, claims about low success rates and possible damage to the embryo, with resulting disability in any viable offspring, have been used to argue against cryopreservation.[2] On the other hand, claims about the reduction of the likelihood of multiple pregnancy resulting from transfer of all embryos generated through IVF, minimization of the stress and cost of repeated egg removals, and possible use of embryos for research purposes have been used to justify cryopreservation.[3] While the claims on both sides deserve further examination, especially in light of serious feminist criticisms of IVF (on which embryo cryopreservation depends) and of contract motherhood (a possible way of using frozen embryos), I shall not attempt here to assess the general justification of cryopreservation, nor of all the issues related to this practice.

Instead, this chapter will analyse and assess what the current practice of cryopreservation in North America assumes and implies about prevailing social and moral concepts of parent-hood, especially fatherhood, and will suggest some tentative con-clusions about parental decision-making with respect to frozen embryos. The possibility of long-term 'banking' of embryos, and recent cases of parental unavailability or disagreement about the disposition of fertilized eggs, raise questions about the nature of relationships to embryos, and about who is entitled to decide what happens to them. Decisions about frozen embryos require the contemplation of a variety of possible alternative outcomes, including prolonged cryopreservation (which only serves to post-pone some moral questions, and even raise others, having to do with posthumous implantation and inter-generational transfer), transfer ('donation') of embryos to the uterus of an unrelated woman, destruction of extra embryos, and use of embryos for further research. While these alternatives are all deserving of evaluation, this discussion will concentrate not on *what* is done to embryos, but rather on *who* gets to decide, and what that suggests about concepts of parenthood and authority over repro-duction.

◆ ◆ ◆
Davis v. Davis: Some Preliminary Concerns

One obvious place to begin discussion of embryo cryopreserva-tion is the Davis case, a landmark example of controversy about frozen embryos. In 1988, nine ova were removed from the ovaries of Mary Sue Davis and fertilized with the sperm of her then husband, Junior Davis. Two of the resulting embryos were placed in Mary Sue Davis's uterus, but no pregnancy resulted. The seven remaining embryos were cryopreserved for possible future attempts at pregnancy. Some months later, however, the couple were divorced, and a legal dispute arose over the disposi-tion of the seven embryos. Mary Sue Davis sought entitlement to have the embryos implanted in her uterus; Junior Davis opposed both her plan to have them implanted and the alternative sug-gestion of anonymous donation to any other woman.

In a Tennessee Circuit Court decision, Judge W. Dale Young opined that the seven embryos were not property, that human life begins at the point of conception, and that 'Mr. and Mrs.

Davis [had] produced human beings, in vitro, to be known as their child or children.' The best interests of these 'children', he concluded, were served by permitting Mrs Davis, their 'Mother' (with a capital M), 'to bring them to term through implantation.' He vested temporary custody of the embryos in Mary Sue Davis, for the purpose of implantation.[4]

This decision is rife with both apparent errors of fact and philosophical confusions, some of which can be described only briefly. First, in his assessment of the status of embryos, Judge Young drew extensively upon the testimony of Jerome Lejeune, described as a world-recognized expert in human genetics. No claim was made that Lejeune possessed any special ethical expertise, yet Young explicitly relied not only on Lejeune's scientific authority but also on his philosophical reasoning. According to Young, Lejeune testified that 'at the moment of conception' a human being with a 'unique personal constitution' has its beginning.[5] Such a claim ignores the extensive body of scientific evidence showing that conception is not a momentary event, but rather a process that takes place over a period of about twenty-four hours.[6] More significantly, however, both Lejeune and Young assumed, without argument, that because the embryos in question are indisputably human, they are also human beings. This latter status, in Young's view, endows them with the same moral status as that possessed by children; the embryos are not property.[7]

But these broad ethical/metaphysical leaps from 'human' to 'human being' and then to 'child' are in no way sanctioned by scientific evidence about fertilization. Young's use of the term 'human' involves an unaware equivocation: he assumes that to say that a living entity is human, a species classification, is equivalent to saying that it is a person, a moral classification. But while the humanness of the embryos is not in question (see Chapter 2), the terms 'human being' and 'child' imply a much wider range of moral entitlements: the sorts of rights attributed to persons.[8] And that status must be argued for, not assumed on the basis of simple genetic constitution. The humanness of a four-celled embryo does not suffice to make it equal in moral standing to, say, a four-year-old child.

Second, there are conceptual ambiguities in talk about the embryo that are either deliberately conflated or accidentally ignored in Young's decision. For example, the judge appears to have rejected the claim made by most of the expert witnesses that the individual cells of the embryo are undifferentiated on the grounds of the quite different claim that a particular embryo can be uniquely differentiated from other embryos on the basis of DNA manipulation.[9] He concluded that '[f]rom fertilization, the cells of a human embryo are differentiated, unique and specialized to the highest degree of distinction',[10] thus confusing the genuine uniqueness of the embryo as a whole with a false claim about the distinctness of the embryo's parts. In fact, up to the eight-cell stage, each single embryonic cell is distinct and totipotent; that is, each one has the capacity to become, separately, an independent individual.[11]

In addition, part of the dispute between Mary Sue Davis and Junior Davis concerned whether or not the embryos were 'alive', or constituted 'life'. While Junior Davis and three of the expert witnesses claimed that the embryos were not life but had the potential for life,[12] Mary Sue Davis argued that 'in order to die [allowing the embryos to die was one solution offered to the Court] one must first live'; if the embryos could suffer a 'passive death, then they must constitute life'.[13]

It seems likely that the two disputing parties disagreed not (or not only) about a matter of fact, but rather about a conceptual question concerning the meaning of the terms 'alive' and 'life'. While Mary Sue Davis seems to have interpreted 'alive' to mean merely 'not dead', Junior Davis took 'alive' to have a more complex meaning having to do with ongoing functioning and development. As David T. Ozar suggests, in a paper pre-dating the Davis case:

> 'life' ... means not only that the organism is not dying, but also that it is able to continue to perform life functions (with or without mechanical assistance) outside of a womb. On this interpretation, the frozen embryo is not viable. For while capable in their frozen state of not dying, these embryos cannot continue to perform life functions, even simple cell divisions, independent of the nutritive and protective environment of a woman's womb.[14]

Young, however, ignored both the ambiguity in the term 'life' and the questionable appropriateness of referring to cryopreserved embryos as 'alive', and ruled that the embryos were simply 'life' from the time of conception.

◆ ◆ ◆

Assessing the Case's Preconceptions

The problems just described are of obvious philosophical interest. But other aspects of the case must be of particular concern to feminists. Junior Davis testified that he opposed Mary Sue Davis's use of the embryos because he did not want to be 'raped of [his] reproductive rights'.[15] Her use of the embryos without his consent would force parenthood on him, he stated, and after Young's decision he complained to the press, 'they are going to force me to become a father against my wishes.'[16] Junior Davis's subsequent appeal of the Young decision claimed that the earlier decision 'was tantamount to the court's deciding that Junior may be required to become a parent against his will, thus denying him the right to control reproduction'.[17] His lawyer stated to the press, 'If we are ever to make men truly equal partners, you can't just say that because she is female, she has greater rights.'[18]

What is astonishing about these comments is their explicit and incendiary appeal to the language and values of two standard feminist political issues: the struggle against nonconsensual sexual activity and the demand for reproductive freedom and rights. Whereas what feminists seek for women is bodily autonomy, the choice of whether or not to be pregnant, and freedom from enforced maternity, what Junior Davis was seeking was to control the destiny of his sperm. His complaints must be heard within the context of a long cultural history in which men have expressed fears about women 'tricking' them by becoming pregnant. There is nothing new about men's worries regarding women's use of their sperm; what is new is the expression of that worry within a co-opted system of feminist language and values.

Do Junior Davis's concerns about the use of the frozen embryos have any moral legitimacy? In an early book on reproductive ethics, philosopher Michael D. Bayles enunciated the principle that 'no one should involuntarily have parental responsibilities', and added that the principle 'prohibits completely involuntary parenthood'.[19] Bayles applies this principle to the defence of a man whose wife becomes pregnant through alternative insemination without his consent; in his view a man's consent to his wife's insemination is an ethical prerequisite for his having parental responsibilities.

However, such a claim unjustifiably disregards the well-being of offspring, rendering them vulnerable by exonerating male partners from parental responsibility whenever they have not explicitly consented to take it on.[20] It would create the possibility of a man's having to assume parental responsibility for only some but not all of the offspring born to his female partner during the partnership itself, so that the interests of some of the children within a family would be better protected than those of others, and it would exacerbate the already serious cultural isolation of mothers.

Moreover, it cannot be assumed that reproductive freedom for men—the absence of which might mean loss of control over donated or commercially supplied sperm—is comparable to women's reproductive freedom, the absence of which means forced reproductive labour and the loss of bodily integrity. Once their sperm has been used to fertilize a woman's ovum, men do not have a right to determine whether a child will be born. Men who want to control their sperm should be careful where they put it, and should pause to think before they provide their sperm for insemination or for *in vitro* fertilization—even with women who are their partners.

Men are therefore entitled to exercise reproductive choice at the time that sperm leaves their body and is conveyed to another location—whether a woman's vagina or a test tube; but there are no grounds for extending male reproductive freedom beyond this point. Hence failure to assign a veto to men like Junior Davis over the use of cryopreserved embryos generated using their sperm is not a violation of their 'reproductive rights', or, more specifically, their right not to reproduce. In application to men, the acknowledgement of a right not to reproduce requires that sexual behaviour be genuinely autonomous, and that sperm donation and sales be genuinely voluntary. It does not require the extension of indefinite control over what is done with sperm after it has been freely provided.

Unfortunately, in the appeal Judge Franks cited both a right to procreate and a right to prevent procreation, and affirmed that Junior Davis had a 'constitutionally protected right not to beget a child where no pregnancy has taken place'.[21] While Franks's decision in fact created no new problems for Mary Sue Davis, who had remarried since the earlier decision and no longer wanted to implant the embryos,[22] and while his views have been lauded by at least one progressive commentator as being both 'sensibl[e]' and 'reasonable',[23] the implications of the decision may be neither sensible nor reasonable.

The judge's reasoning seems to imply an entitlement of donors and vendors[24] of both sperm and eggs to retain ongoing control over their gametes, even after the gametes have, by the donors' or vendors' choice, been removed from their bodies. Similarly, a recent report on ethical issues relating to new reproductive technologies advocates that a woman 'should have control over what happens to her eggs since they are her eggs'.[25] The suggestion is that donors and vendors of either sex should be able to determine whether their donated gametes are used for research, to produce a pregnancy in recipients, or for both these purposes. The crucial question, however, is whether one's gametes remain one's own after one has donated or sold them for use by another person, or, particularly, after they have been combined with another's gametes to form new embryos.

John A. Robertson, a professor of law who testified as an expert witness at the original *Davis v. Davis* hearing, claims:

> Just as one 'owns' their own body and its parts vis-à-vis other persons, so gametes are owned by the parties providing them or their transferees. It follows that the embryo resulting from the fusion of gametes from two persons is owned by the persons providing the gametes. The gamete providers' wishes should control over the wishes of other parties, at least until they transfer that authority to other persons.[26]

But when should the transfer of authority be considered to have occurred? I suggest that gametes that have been donated or sold are no longer 'one's own' in terms of any entitlement to determine their disposition, and that embryos resulting from the combination of one's own gametes with those of another are not simply 'one's own' in the same way the original gametes were. While it may be important to ensure that prospective donors have the initial choice whether their gametes are used for inducing pregnancy (via alternative insemination or *in vitro* fertilization) or for research, there is no justification for further rights of disposal such as the entitlement to determine whether one's gametes go to a person of the same race or sexual orientation as the donor, or to decide that the gametes may not be used at all if a recipient of a donor's preferred sort is not available. The fact that sperm or eggs once originated from a particular individual does not give that individual an entitlement to impose his or her agenda or prejudices on the disposition of the gametes once they have been provided to an individual or institution for further use, and particularly once they have been combined with the gametes of another person.

Franks's assertion of a 'right not to beget a child where no pregnancy has taken place' could have additional undesirable implications for donors' control over embryos produced from their gametes. It might, for example, imply a right to withdraw embryos after a couple had donated them to another woman. Thus in a 1987 paper, Robertson argued that '[a] person's interest in having or avoiding biologic heirs also supports the gamete providers' authority over embryos formed from their gametes.'[27] According to Robertson, this authority should extend even past one's own life: he seeks to recognize the 'procreative rights of the gamete providers to reproduce after death'.[28] But once a body part or product has, by one's own choice, left one's body, there are morally justifiable limits on the degree of control one may have over it. As for the recipient, before the donation she is not, of course, entitled to the embryos, but after the donation she has acquired an entitlement to them, which ought not to be withdrawable by the original donors.

In the appeal, Judge Franks reasoned that just as it would be 'repugnant and offensive' to order Mary Sue Davis to be implanted with the embryos if she chose not to be, so also 'it would be equally repugnant to order Junior to bear the psychological, if not the legal, consequences of paternity against his will'.[29] But this judgement betrays an erroneous assumption about analogies between reproduction in women and reproduction in men. If the embryos were successfully implanted in Mary Sue Davis against the will of her former husband, Junior Davis would have become a parent only in the minimal genetic sense of being biologically related to the resulting offspring. Bayles's reason for prohibiting 'involuntary parenthood' is that 'parental responsibilities and rights are significant burdens and privileges which can greatly affect people's lives'.[30] While this is true in most cases, and is certainly true for all women who actually gestate and bear children, it need not have applied to Junior Davis, since Mary Sue Davis was quite willing to raise the child alone, without his support or involvement. She did not seek to impose on her former spouse any legal, economic, or social responsibilities for any children that might result from implantation. So Junior Davis would have had a choice not to be a 'social parent' to any children Mary Sue Davis might bear as a result of implantation of the embryos.

Hence a concern for Junior Davis in the event of his former wife's successful pregnancy would have to be founded upon a concern for the significance of genetic parenthood, and biological, not emotional or social, attachment to one's offspring. Robertson recognizes that '[w]hether an unwanted but unidentified biological link is sufficient to ground a right will depend upon the social and psychological significance which individuals and society place on the existence of lineal descendants when anonymity and no rearing obligations exist'.[31] Predictably, however, he takes for granted 'the reproductive and personal significance of potential biologic offspring for the gamete source', and argues that '[a] person's interest in avoiding biologic heirs, even if they do not rear them, is significant and deserves respect.'[32] The argument is that there is an important value simply in avoiding what Janice Raymond has called 'ejaculatory fatherhood':[33]

[T]he very concept of fatherhood is being extended to include sperm donors, as if by virtue of ejaculation alone a man becomes a father.... What we see in repeated litigation involving new reproductive technologies is another version of what Mary O'Brien has termed 'ejaculatory politics'. Ejaculation doth a father make. Ejaculation confers father-right. This is not mere metaphor but grossly material if you will.[34]

But of course it cannot merely be assumed, without argument, that avoiding a simple genetic connection with offspring has any moral significance at all, let alone a significance so great as to justify refusing to release embryos to the woman who underwent extensive medical interventions in order for them to be generated.

The only evidence of such a significance that is provided in the proceedings of the hearing—and the only evidence that is ever provided in debates where biological ties are taken to be significant—consists of statements about feelings concerning the value of a biological tie. In the Davis case, we have to estimate the significance of the feelings Junior Davis would experience as a result merely of knowing that a child or children existed who were genetically related to him. These feelings were described by Robertson, in the initial hearings, as 'the traumatic psychological burdens of being forced to be a parent against his will'.[35] But again, the only respect in which Junior Davis would be forced to be a parent would be as a genetic father, not a social one. The burdens here are not the burdens that would result if children were born whom Junior Davis came to know, and from whom he was then forcibly separated. Junior Davis is not complaining of the potential pain of separation from his children; he is complaining of the potential pain of being biologically related to children whom, if he so chooses, he need never know or even see. So the situation here is not in any way analogous to that of a woman who gestates a foetus, subsequently surrenders or is forced to surrender the infant for adoption, and then regrets having done so.

Nevertheless, this fact was apparently no comfort to Junior Davis, who testified that he did not want 'a child produced [through implantation of the embryos in Mary Sue Davis's uterus] to live in a single-parent situation'. His argument was that he himself had suffered because of the divorce of his parents; he grew up in a boys' home and experienced 'despair because there was no natural bond with his parents':

He strongly and sincerely insists that because of his poor relationship with his own parents he strenuously objects to bringing a child into the world who would suffer the same or a similar experience without any opportunity on his part to bond with his child.[36]

Given that Mary Sue Davis was committed to a permanent relationship with her child, Junior Davis was assuming that his own absence from the life of his offspring would be a major impediment to its well-being. So on the one hand, he was trying to prevent the existence of his genetic children; on the other hand, he was also worrying about their possible suffering without his presence in their lives. Since of course Junior Davis's life could have been just as miserable if his birth family had been intact, he seemed to believe in an almost-mystical power of genetic fathers, or of himself in particular as a genetic father, to prevent children from experiencing despair.

Junior Davis further testified that if a child were born to Mary Sue Davis as a result of the implantation of the embryos, he would try to develop a relationship with it; he would actively seek both to support and to gain custody of the child. Hence it must be assumed that Junior Davis saw some potential value in supporting and caring for his potential future children. This value is difficult to reconcile with the notion of 'traumatic psychological burdens'.

In his testimony for the Davis case, Robertson claimed that in the 'balancing of the equities', Junior Davis would be more injured by being made to become a (genetic) parent than Mary Sue would be injured by being prevented from implanting the embryos.[37] How was this assessment of harms arrived at? In his earlier paper, Robertson states, without argument, that what he regards as the interest of 'persons' (sex unspecified) 'in avoiding biologic heirs, even if they do not rear them, is significant and deserves respect'.[38] Moreover, since achieving pregnancy is uncertain with the use of frozen embryos, Robertson assesses the destruction of embryos as at most a 'purely psychological' loss for the individuals who provided the gametes for them,[39] and only because the individuals have 'an important procreative interest at stake'. The value of the loss should, in his view, be estimated simply on the basis of the 'cost of creating the embryo'.[40] This loss, then, is said to be more than balanced by Junior Davis's interest in avoiding biologic heirs.

Robertson does not specify what the 'cost of creating the embryo' includes. But, given his assumption in the Davis case that Mary Sue Davis could 'apparently successfully participate in the IVF program with another partner in the future',[41] he is likely not including the often-hidden costs involved in the creation of embryos. In the Davis case these costs would include what Judge Young himself, on the basis of Mary Sue Davis's testimony, called 'the painful, physically trying, emotionally and mentally taxing ordeals she endured to participate in the [IVF] program'.[42] After an earlier history of five tubal pregnancies, resulting in the rupture of one fallopian tube and the tying of the other,[43] Mary Sue Davis underwent extensive hormonal priming on many occasions (the number is unspecified) to prod her body into producing multiple ova. She was subjected to twenty-one aspirations of eggs, and the transfer of fourteen embryos during six attempts, at a cost of between $4,000 and $6,000 for each attempt.[44] The 'costs' for Mary Sue Davis's loss of the embryos should also include the further physical and psychological pain, medical risks, and economic expenses associated with additional attempts to generate and remove ova, then fertilize and implant them. Her physician, Ray King, testified that there were no guarantees that Mary Sue Davis could ever produce another usable egg, although he optimistically estimated her chance of pregnancy using the existing embryos at 52 per cent.[45] Finally, in assessing the 'balance of equities' between Mary Sue Davis and Junior Davis, the likelihood of difficulties and lack of success in the generation of new embryos in future IVF attempts must also be factored in.

So, whatever the perceived psychological 'burdens' to Junior Davis of becoming merely a genetic parent, it is impossible to accept Robertson's claim that they would have been greater than the very real material burdens to Mary Sue Davis if she were to lose the embryos she was seeking.

♦ ♦ ♦

Parenthood and Children

The value of the Davis case for a feminist assessment of the cryopreservation of 'spare' human embryos lies in the opportunity it provides to raise more general questions—questions about the status of embryos themselves and their relationship to the persons whose gametes engender them. Do embryos have a moral status that is independent of their location or of their connection to persons? This question has, of course, been the subject of extensive exploration in connection with the abortion controversy. Feminists have argued that in instances of unwanted pregnancy, the embryo or foetus cannot be treated in moral isolation, but must be viewed within the context of its corporeal location in a woman's body. But context might seem less important when embryos exist outside of a woman's body. Thus Ozar writes that, in abortion decisions,

> the state's interest in protecting [embryos'] potential life could not outweigh the fundamental constitutional right of a woman to control her own body. But in the case of frozen embryos, no woman is involved, and thus no woman's right to control her body.[46]

So, according to Ozar, obligations to cryopreserved embryos can be considered independently of concern for women. With this assumption about the radical independence of fertilized eggs, cryopreserved embryos are often seen as possessions, objects, instruments, or even 'unclaimed luggage'.[47] Robertson is most enthusiastic about this approach, regarding embryos as potential investments and bankable capital:

> Cryopreservation of embryos will ... facilitate such novel means of family formation as egg and embryo donation and gestational surrogacy.... Couples not ready to form a family might bank embryos as insurance against future sterility or age-related birth defects.[48]

He adds, 'Widespread embryo banking with shipment to distant points will increase the choice of prospective recipients'.[49] In this co-optation of the feminist concept of reproductive freedom, procreative choice becomes consumer choice, facilitated by market forces that make available and regulate the buying and selling of cryopreserved embryos.

The apparent obverse of this reification and commodification of embryos as purchasable instruments for furthering individual goals is the growing fetishization of fertilized eggs. The *Merriam-Webster Dictionary* defines 'fetish' in part as 'an object ... believed to have magical powers (as in curing disease); an object of unreasoning devotion or concern.' The transformation of the embryo into a fetish in this sense is evidenced in a practice reported by Andrea Bonnicksen:

> Patients may develop attachments to their embryos during regular IVF, as indicated by their naming the embryos, asking for the petri dishes in which the embryos were fertilized as mementoes, acting and feeling pregnant after the embryos are transferred to their uteruses, and mourning the embryos' loss if they do not implant.[50]

Indeed, Mary Sue Davis testified at the first hearing that she regarded herself as the mother of the embryos, that she felt an attachment to them, and viewed them as children.[51] While concern for and attention to these women's experiences is of course essential to any feminist analysis of embryo cryopreservation and IVF, their perceptions of the embryos must also be assessed critically in terms of both the pronatalist attitudes that give rise to and may even be intensified by infertility treatments and the promotion of embryos as child-substitutes. For example, some clinics apparently cater to this fetishization of embryos by providing death rituals for the disposal of the embryos;[52] and Robertson notes that donating embryos for research may provide 'meaning' for couples.[53] Through these processes, the embryo appears to become a child, or a child substitute, in the eyes of the gamete providers. Bonnicksen adds:

> To clients, the embryo symbolizes hope and potential parenthood. It affirms the wife's femininity, the husband's masculinity, and the couple's potency. It is a powerful symbol with which clients establish emotional connections. It may be the closest thing to parenthood the wife and husband experience.[54]

Within the lived relationships of gamete providers to their embryos, Judge Young's decision that the seven embryos are children in need of judicious care appears to acquire its experiential validation. Young saw his role as that of *parens patriae*, literally 'father of the fatherland', seeking the best interests of the children whose well-being he believed was in question,[55] and he cited Lejeune's view that his decision was comparable to

Solomon's biblical assessment of the true claimant to mother-hood in a dispute over a baby.[56]

But the cryopreserved embryo is a frail vessel indeed for bear-ing such emotional, social, and moral weight. What does it say about adult/child relationships in North American culture when, for some adults, yearning over an embryo in a dish is an instan-tiation of parenthood? What does it say about the status of chil-dren within North American culture when care and love are lavished on a four-celled fertilized egg, and four-year-olds go hungry? Certainly embryos have the advantage of being, in some ways, both easier to handle and less demanding than children; though it is not cheap to maintain an embryo in a cryopreserved state, it is probably less expensive and certainly less demanding on the 'parents' than feeding, clothing, and educating a toddler. And while commentators ranging from Judge Young to George Annas may reject the notion that embryos can be owned, still it is easier to give away, sell, or destroy an unwanted embryo than to dispose of an unwanted child. The more commonly embryos are regarded as being physically and morally independent of women's bodies, the more they come to be regarded as pur-chasable and undemanding child-substitutes for the 'parents' that generate them.

♦ ♦ ♦
Decision-Making About Embryos

Do 'ejaculatory fathers' (i.e., sperm providers) have any moral rights to control over embryos? How should competing claims from women and men about reproductive rights with respect to embryos be handled?

Ideally, of course, gamete contributors would agree about the disposition of resulting embryos, and would make prior arrange-ments concerning the treatment of the embryos and the role and authority of the storage facility in the event of the death of one or both, divorce or separation in the case of married couples, or a decision not to undergo future implantations. Their decisions should include assessment of the various alternatives of discarding, donating for research, or giving the embryos to another women— without being entitled to specify which woman, or type of woman, is entitled to have the embryos. But in the absence of such agreements, a decision like that of the appeal court in the Davis case, to give joint control of the embryos to the disputing parties, simply assigns a *de facto* veto to men like Junior Davis who seek to prevent their ex-partners from gestating the embryos.

Should their location make a difference to our assessment of who should make decisions about embryos? In questions of abortion, the location of the embryo or foetus in a woman's body is crucially significant to her entitlement to decide whether to continue the pregnancy. In the case of cryopreserved embryos, the fact that the embryos are outside a woman's body does not necessarily make her interest in decision-making about them less than it is when she is pregnant, for the location of the embryos, in a petri dish or freezer, does not make them independent of a woman's body. Ozar's claim that with laboratory-produced embryos 'no woman is involved' is false. These insouciant words betray the error, egregiously shared by the embryo-fetishists, of regarding the embryo as a sort of technological *Ding an sich*, a being existing independently of any woman's body. For at least one woman is crucially involved: that is, the woman who pro-vides the eggs from which the embryos are produced, and into whose uterus the embryos may be implanted. The involvement of this woman gives her the entitlement to decide what happens to the embryos.

But it is important to be clear about why her involvement con-fers this entitlement. Janice Raymond straightforwardly appeals to the woman's 'greater contribution to the embryos, ... her repeated attempts at implantation and gestation, and considerable bodily investment'.[57] Robertson, by contrast, rejects what he calls the '"sweat equity" position that always favors the woman's decision because she has put more effort into production of the embryo, having undergone ovarian stimulation and surgical retrieval of eggs'. The reason, he writes, is that

> [g]reat differences in physical burdens do not require that divorc-ing mothers always receive custody of children. Moreover, the dif-ference in bodily burdens between the man and woman in IVF is not so great ... that it should automatically determine decisional authority over resulting embryos.[58]

Yet Robertson's belief that embryos are property is inconsistent with his assumption that custody decisions about children are relevant to decision-making about embryos. As an advocate of the view that embryos are not persons, and not comparable to children, Robertson is the last person who should appeal to legal decisions about the custody of children.

Moreover, the argument that the difference in 'bodily burdens' between women and men in IVF is insufficiently great to justify assigning decision-making to the woman is based on an implausible premise. There is presumably little burden involved in masturbating to produce sperm, whereas the physical and psychological burdens of hormonal treatments, aspiration of eggs, and implantation can be almost overwhelming.[59] Judge Young claimed that Junior Davis

> spent many anxious hours, early in the morning and late at night, waiting at the hospital while Mrs. Davis underwent the aspiration and implant procedures and ... he spent many anxious hours, as a prospective Father, awaiting word as to whether he would be a Father.[60]

But if the man's anxiety and uncertainty over the possible outcome of IVF are evidence of his 'burdens', they hardly compare to the physical and emotional stress on the person whose body is the site of the interventions, and who experiences even greater stress about their consequences.

Nevertheless, the argument for assigning decisional authority for embryos to the woman should rest not on an appeal to the woman's 'investment', a claim that too readily buys into the view of embryos as capital. And it should not, or not only, rest on the woman's clearly greater prior burdens. Nor should it involve, as Janice Raymond tentatively suggests, an appeal to the woman's 'right to what issues from her body',[61] since in the process of *in vitro* fertilization, ova, not embryos, are what issue from her body, and the sperm provider can equally argue that the sperm that fertilized the ova issued from his body. Instead, the argument should be future-oriented. What the Davis case shows is that the justification for assigning decisional authority over embryos to the woman, in cases of dispute, is to reduce the likelihood that she will have to undertake the burdens of IVF in the future. Giving the embryos to the woman for implantation means that she may be able to avoid undergoing further treatment with massive amounts of hormones and the removal of more ova.

The justification for giving the embryos to the woman can also be demonstrated by imagining the consequences if the intentions of the disputing parties in the Davis case had been reversed. Suppose, that is, that the sperm provider wanted the embryos, and his ex-wife, the woman who supplied the ova, wanted them to be destroyed. What are the implications if the woman's decision is not determinative? It seems that the embryos would be given to the man. But since he cannot gestate them, and since, as even Judge Young recognized, it would be wrong to *force* the woman to gestate them against her will, it would seem to be necessary for the man to find another woman who is willing, for payment or other reasons, to gestate them. In other words, the claim that it is appropriate to 'give the embryos to the sperm provider', against the will of the woman who provided the eggs, requires in its practical implications that some form of contract motherhood, whether commercial or 'altruistic', be endorsed.

The arguments against this practice are discussed at length in Chapter 7. Here it is sufficient to note that the complex and serious problems relating to the recruitment and possible exploitation of women in contract motherhood, and to later custody of any resulting children, militate against its endorsement.[62] While commercial 'surrogacy' raises obvious problems with respect to the misuse of women's bodies and the sale of infants, so-called 'altruistic surrogacy' is also of questionable moral value, since it is not clear to what degree the 'surrogates' participate willingly, and their infants, while not purchased, are handed over like objects to the recipients, without concern for the infants' own interests. To provide embryos to men in situations like that of Junior Davis would amount to giving tacit state approval to some forms of contract motherhood. In view of the very strong reasons against state endorsement of this practice, the embryos ought not to be provided to the man—who in most cases is still fertile and capable of future procreation through heterosexual intercourse or donor insemination.

Suppose, however, that Junior Davis had become infertile shortly after the end of his marriage. If Mary Sue Davis did not want the embryos implanted in her uterus, should they not then be given to Junior, even against her will?[63] Such a proposal assumes without argument that the significance to Junior of having his own genetically related children is greater than the significance to his future female partner of having her own genetically related offspring. For if his female partner gestates the embryos of Mary Sue and Junior Davis, any resulting child will, of course, not be genetically related to her. Moreover, the woman's chances of a successful pregnancy and birth, with few medical interventions, would be much higher if she were to be inseminated with donor sperm. To give the embryos to Junior, therefore, is to assume the willingness of his future female partner to choose the much riskier reproductive path of embryo implantation. Indeed, it is implicitly to *endorse* that path: to say that it is preferable for Junior's future female partner to attempt gestation of unrelated embryos, with all the attendant risks of such a pregnancy, than to go through a low-risk pregnancy initiated through alternative insemination.

Nevertheless, it is imaginable that Junior's future female partner herself might have certain problems with infertility: for example, blocked fallopian tubes, the classic indication for IVF. In this circumstance, it might be argued, Junior should be given the embryos for precisely the sort of reason outlined earlier for giving them to Mary Sue Davis: to reduce the likelihood that his female partner will have to undergo a future course of hormonal priming and egg withdrawal. Notice, however, that this proposal too assumes, without evidence, that the female partner will choose gestation of unrelated embryos over gestation of her own: that is, it assumes her consent to a procedure designed to guarantee Junior's genetic connection to the offspring, but not her own. Indeed, it assumes, without evidence, her willingness to be a mother in any way. But, aside from begging the question of the female partner's choice with respect to her reproductive activities, and perhaps even putting pressure on her to engage in such activities, giving the embryos to Junior Davis implies that the physical and emotional pain and medical risks suffered by Mary Sue Davis should be used to benefit another woman, even against Mary Sue's will. While it would perhaps be good of Mary Sue Davis to surrender the embryos—providing that Junior

Davis's future female partner wanted to gestate them—it would be unfair to compel her to do so. Giving the embryos to Junior Davis, against Mary Sue's will, would be disrespectful of her suffering, and would inappropriately treat Mary Sue's IVF ordeal as a means to another woman's procreative end.

I conclude, therefore, that giving the embryos to Junior Davis, against Mary Sue Davis's will, has one or more of the following consequences: endorsing contract motherhood; presuming upon the procreative choices of and placing pronatalist pressure on Junior Davis's future female partner; affirming the value of embryo implantation over donor insemination; or using one woman's medical treatment as a means to another woman's reproductive goals. Since all of these consequences are undesirable and undeserving of state endorsement, the embryos should not be given to Junior Davis against Mary Sue's will, even when Mary Sue does not intend to have them implanted in her own uterus.

In cases of disagreement about the disposition of cryopreserved embryos, therefore, joint decision-making gives an effective veto to the man, and giving the embryos to the man entails the *de facto* endorsement of several morally questionable assumptions and practices. Moreover, under these conditions, the woman faces the burden of undergoing IVF once again. Hence decision-making about cryopreserved embryos should, in cases of disagreement, be assigned to the woman, who is entitled to choose whether or not they will be implanted in her uterus.

◆ ◆ ◆
Some Final Comments

The current terms of much of the debate about the disposition of embryos, a debate that often includes appeals to right-wing views about 'maternal feelings' and women's proper reproductive role, lends some credence to the radical feminist speculation that control of reproduction is central to the preservation of patriarchy.[64] '[New reproductive] technologies focus medical, legal, and media attention on the status and rights of fetuses and men while rendering the status and rights of women at best incidental and at worst invisible.'[65] The dispute over a four-celled organism is at least as much a debate about reproductive control of women. The message of Junior Davis, of his lawyers, supporters, and approving commentators, and of the appeals court judge, is that even after a woman undergoes the arduous and usually unrewarding processes of IVF, she can and should be prevented from deciding her reproductive future. The subtext appears to be fear—fear of the spectre of sperm theft, and fear that a man's loss of control over his sperm is the loss of control over his life. The Davis case illustrates not only the ongoing fetishization of embryos, but also the fetishization of sperm, and an uncritical yet morally problematic equation of ejaculation with fatherhood, and embryos with babies.

Notes

I wish to thank Joan Callahan for her challenging and provocative comments on an earlier draft of this chapter, and Jennifer Parks, whose work on the issue of 'respect' for embryos helped me to think through some of the problems discussed here.

[1] I use the term 'embryo' throughout this chapter to refer generally to entities produced through the combination of a human ovum with a human sperm, from fertilization to approximately two months' development. The embryos discussed in this chapter have usually developed only for a few days, although in another sense they can be much 'older', by virtue of having been cryopreserved, in a very early stage of development, and maintained for months and even years in that state. I chose not to use the term 'pre-embryo', which is advocated by some writers as appropriate for the post-conception entity up to fourteen days of development, when the primitive streak first appears. (The best example of use of this term is in the anthology *Embryo Experimentation*, edited by Peter Singer et al. [New York: Cambridge University Press, 1990].) In my view, the arguments for making a terminological distinction of this sort are not convincing.

[2] Hans O. Tiefel, 'Human In Vitro Fertilization: A Conservative View', in Richard T. Hull, ed., *Ethical Issues in the New Reproductive Technologies* (Belmont, CA: Wadsworth, 1990), 129.

[3] American Fertility Society Ethics Committee, 'Ethical Considerations of the New Reproductive Technologies', *Fertility and Sterility* 42, 3, Supplement 1 (September 1986): 53S.

[4] *Davis v. Davis v. King* E-14496 (Fifth Jud. Ct. Tennessee), Young, Judge (1989), 1-2. To be consistent with his own decision, Judge Young should have added that, since grand multiple pregnancies are more

dangerous for foetuses, the embryos should be implanted in stages and gestated during several pregnancies, in order to provide them with a greater chance of surviving to birth.

[5]*Davis v. Davis v. King*, 28.

[6]Karen Dawson, 'Fertilization and Moral Status: A Scientific Perspective', in *Embryo Experimentation*, 43.

[7]*Davis v. Davis v. King*, 10.

[8]Mary Anne Warren, 'On the Moral and Legal Status of Abortion', in Joel Feinberg, ed., *The Problem of Abortion* (Belmont, CA: Wadsworth, 1984), 110.

[9]*Davis v. Davis v. King*, 9.

[10]Ibid., 1.

[11]Helga Kuhse and Peter Singer, 'Individuals, Humans and Persons: The Issue of Moral Status', in *Embryo Experimentation*, 67.

[12]*Davis v. Davis v. King*, 5 and 10.

[13]Ibid., 9 and 27.

[14]David T. Ozar, 'The Case Against Thawing Unused Frozen Embryos', *Hastings Center Report* 15, 4 (1985): 8.

[15]*Davis v. Davis v. King*, 21.

[16]Ronald Smothers, 'Woman Given Custody in Embryo Case', *New York Times* (22 Sept. 1989).

[17]*Davis v. Davis* WL 130807 (Tenn. App.), Franks, Judge (1990): 2.

[18]Quoted in Janice Raymond, 'Of Ice and Men: The Big Chill Over Women's Reproductive Rights', *Issues in Reproductive and Genetic Engineering: Journal of International Feminist Analysis* 3, 1 (1990): 49.

[19]Michael D. Bayles, *Reproductive Ethics* (Englewood Cliffs, NJ: Prentice-Hall, 1984), 16.

[20]See Christine Overall, *Ethics and Human Reproduction: A Feminist Analysis* (Boston: Allen and Unwin, 1987), 183.

[21]*Davis v. Davis*, 2.

[22]Ibid., 4.

[23]George J. Annas, 'Crazy Making: Embryos and Gestational Mothers', *Hastings Center Report* 21, 1 (1991): 35 and 36.

[24]The word 'donors' is usually used misleadingly to refer not only to men who donate sperm but also to those who sell it. For the latter, the term 'vendor' is more appropriate.

[25]Combined Ethics Committee of the Canadian Fertility and Andrology Society and the Society of Obstetricians and Gynaecologists of Canada,

Ethical Considerations of the New Reproductive Technologies (Toronto, 1990), 40.

[26]John A. Robertson, 'Embryos, Families, and Procreative Liberty: The Legal Structure of the New Reproduction', *Southern California Law Review* 59, 5 (1986): 976.

[27]John A. Robertson, 'Ethical and Legal Issues in Cryopreservation of Human Embryos', *Fertility and Sterility* 47, 3 (1987): 373.

[28]Ibid., 374. By contrast, the American Fertility Society Ethics Committee recommends that 'Storage should be continued only as long as the normal reproductive span of the egg donor or only as long as the original objective of the storage procedure is in force' ('Ethical Considerations', 55).

[29]*Davis v. Davis*, 3.

[30]Bayles, *Reproductive Ethics*, 16.

[31]Robertson, 'Embryos, Families, and Procreative Liberty', 979.

[32]Robertson, 'Ethical and Legal Issues', 376.

[33]Raymond, 'Of Ice and Men', 47.

[34]Ibid., 46 and 48.

[35]*Davis v. Davis v. King*, 22.

[36]Ibid., 21.

[37]Ibid., 22-3.

[38]Robertson, 'Ethical and Legal Issues', 376.

[39]Ibid., 379.

[40]Ibid., 380.

[41]*Davis v. Davis v. King*, 23.

[42]Ibid., 27.

[43]Raymond, 'Of Ice and Men', 46.

[44]*Davis v. Davis v. King*, 24 and 27.

[45]Ibid., 24.

[46]Ozar, 'The Case Against Thawing', 8.

[47]George J. Annas, 'Redefining Parenthood and Protecting Embryos: Why We Need New Laws', *Hastings Center Report* 14, 5 (1984): 51.

[48]Robertson, 'Ethical and Legal Issues', 371.

[49]Ibid., 376.

[50]Andrea L. Bonnicksen, 'Embryo Freezing: Ethical Issues in the Clinical Setting', *Hastings Center Report* 18, 6 (1988): 27.

[51]*Davis v. Davis v. King*, 27.

[52]Bonnicksen, 'Embryo Freezing', 29.

[53]Robertson, 'Ethical and Legal Issues', 378.

[54]Bonnicksen, 'Embryo Freezing', 29.

[55]*Davis v. Davis v. King*, 11.

[56]Ibid., 31.

[57]Raymond, 'Of Ice and Men', 46.

[58]John A. Robertson, 'Resolving Disputes Over Frozen Embryos', *Hastings Center Report* 19, 6 (1989): 7.

[59]Linda S. Williams, 'No Relief Until the End: The Physical and Emotional Costs of In Vitro Fertilization', in Christine Overall, ed., *The Future of Human Reproduction* (Toronto: Women's Press, 1989).

[60]*Davis v. Davis v. King*, 3.

[61]Raymond, 'Of Mice and Men', 46.

[62]See 'The Case Against the Legalization of Contract Motherhood', chapter 7 below.

[63]I am grateful to Joan Callahan for suggesting this scenario and the next one.

[64]Jalna Hanmer, 'Transforming Consciousness: Women and the New Reproductive Technologies', in Gena Corea et al., eds, *Man-Made Women: How New Reproductive Technologies Affect Women* (Bloomington: Indiana University Press, 1987).

[65]Raymond, 'Of Ice and Men', 45.

Chapter 6

♦ ♦ ♦

The Co-optation of Feminist Values in Defence of Reproductive Engineering: A Case Study

In current discussions of reproductive technologies and their social uses, feminist ideals—the moral language of 'rights', 'freedom', 'consent', 'oppression', 'sexism', 'choice', and 'control'—are being used by non-feminists and even by some self-described feminists to serve in the defence of misogynist reproductive practices. This co-optation of feminist moral language is a fairly widespread phenomenon, also occurring in other areas of current social-policy formation (for example, the disposition of cryopreserved embryos, discussed in Chapter 5, and the so-called 'fathers' rights' movement[1]). The language can sometimes be misleading, perhaps even to feminist activists and theorists who are using the same terms. To illustrate this misappropriation of feminist discourse, this chapter discusses and evaluates one conspicuous example, a paper by an anti-feminist woman on the topic of so-called 'surrogacy' contracts.

The paper, entitled 'FOR the Legitimacy of Surrogate Contracts', by H[arriet] E. Baber, appears in an anthology published in late 1987,[2] which contains papers both with and without a feminist orientation. Baber's paper presents several criticisms of some standard arguments against 'surrogacy' arrangements, some

105

of which have been advanced by feminists. Baber couches her criticisms partly within the language most of us would associate with feminism. She claims, for example, that 'standard objections to the practice embody unacceptable and empirically unwarranted sexist assumptions'.[3] On examination, Baber's paper provides a compelling illustration of the co-optation of feminist language and ideals.[4]

Baber addresses what she takes to be three standard arguments against 'surrogacy' contracts (although she does not cite any sources for these arguments).

◆ ◆ ◆

Argument I

Baber calls this argument 'the right to control one's body':[5]

1. No one can have a right over another person's body or any part of another person's body.

2. A foetus is a part of its mother's body.

3. Therefore no one but its mother can have any right over it.

4. Surrogate-parenting contracts purport to confer parental rights over the foetus to other persons.

5. Since (by 3) this cannot be done, all such contracts are null and void.

Although Baber indicates that she is herself an opponent of abortion and does not accept that the foetus is a part of the pregnant woman's body,[6] she grants for the sake of the argument that the foetus is a part of the woman's body, and that no one can have a right to a current part of another person's body. But she then argues that it is necessary to distinguish between the time at which rights are acquired and the time at which the agreement to transfer rights is made: 'a person may enter into an agreement transferring a right to another party prior to the time at which the other party is in the position to acquire that right'.[7] Hence, according to Baber, in a 'surrogacy' contract the hired woman agrees to transfer parental rights to a child at a point when the hiring father is not yet in a position to acquire the rights, because the child is still a part of the mother's body. The hiring father then acquires the rights to the child many months later, when the new baby is no longer a part of its mother's body. Thus in Baber's view we need not regard the 'surrogacy' arrangement as immoral on the grounds that it usurps the woman's control over her body, since what the purchaser acquires are not rights over the woman's body or parts of it, but rather rights to the baby only after it is born.

Why, Baber asks, 'if a woman may relinquish her parental rights over her child after its birth, ... may she not do so prior to its birth, or, indeed, prior to its conception? Why should we not regard "surrogate" parent contracts as pre-natal adoption papers?'[8] The answer is that even if 'surrogacy' can be compared to adoption (and there may be important ways in which it cannot), then surely at least some of the minimal protections adhering to adoption practices should be applied. Most notably, the biological mother ordinarily has a period of grace after the birth in which she has the opportunity to reconsider the decision to give up the child. Such an opportunity is entirely absent from 'surrogacy' arrangements.

This limitation is typical of the restrictions imposed on women by 'surrogacy' contracts—restrictions of which Baber seems unaware. Thus while Baber grants, for the sake of argument, the feminist claim that no one has a right to use or control part of another person's body, she fails to acknowledge the ways in which 'surrogacy' contracts violate women's bodily autonomy. Such contracts ordinarily put serious limitations on the woman's work and leisure activities, her travel, her medical care, even her sexual activities. They may specify that she must undergo medical tests such as amniocentesis and ultrasound, and may require her both to have an abortion if the foetus is judged by the hiring father to be abnormal, and to waive her right to an abortion if it is not. Furthermore, both those who set up the contracts and those who want to see them enforced intend the contract to be one from which the woman is not permitted to withdraw. In this respect, then, the 'surrogacy' arrangement is like a form of indentured serfdom: for the duration of the contract, the woman's body is not her own. She cannot back out of the contract, and must serve her time, regardless of subsequent doubts or problems she may experience. For these reasons, then, and contrary to Baber's claims, the 'surrogacy' arrangement *does* seriously undermine the woman's right to control her own body.[9]

◆ ◆ ◆

Argument II

According to Baber, this argument, which she entitles 'Surrogate Parenting is Baby-Selling', claims that 'surrogacy' is morally wrong because 'the practice is detrimental to the interest of babies, who are bought and sold like commodities and otherwise treated as mere means to the ends of their natural and adoptive parents'.

Baber's response to this argument involves several counter-claims. First, she says, there are no consequentialist reasons to suggest that it is in the interest of children to be raised by their natural parents, since 'anyone with functioning reproductive organs and a willing partner can make a baby', and there is no evidence that natural parents, even mothers, bond with their offspring or are 'imbued with special intuitions and hormones which guarantee that they will subsequently act in their children's best interests'.[10] Moreover, she says, even if children in 'surrogacy' arrangements are in fact conceived as means to 'their mothers' self-interested ends', this is no different from what happens in other pregnancies. For example, '[w]omen get pregnant in order to bully reluctant partners into matrimony and to shore up faltering marriages'.[11] According to Baber, conceiving a child to further some ulterior end is not morally wrong *per se*, and so it is not wrong in the case of 'surrogacy'.

Presumably the views on bonding and motherhood cited by Baber are examples of what she regards as 'unacceptable and empirically unwarranted sexist assumptions'.[12] The problem is that Baber unjustifiably appropriates the feminist concept of sexism—unwarranted discrimination on grounds of sex—to describe feminist attempts to call attention to the value and importance of mothers. Admittedly, feminists must be careful of what we say about mothering. To accord a special, *a priori* mothering capacity to biological mothers[13] is to denigrate the excellent nurturing that adoptive mothers and other caregivers have for eons provided to children not biologically related to them. In addition, it serves to emphasize and validate the patriarchal view of the importance of a genetic link with one's offspring. The significance of such a link is open to serious question: it cannot merely be assumed, but must be justified.[14] Furthermore, the innatist argument that women are biologically best suited to rear children not only is unfounded but also can be used against women to reinforce traditional maternal roles.

Nevertheless, some feminists would argue that there is a special connection of the mother with her biological child by virtue of the reproductive labour she has expended in creating the child.[15] That reproductive labour obviates the claim made, for example, by Judge Sorkow in the first Baby M decision, that the baby in a 'surrogacy' contract already belongs to the biological father, and that the mother is simply returning what he already owns.[16] Clearly, the reproductive contributions made by the male and the female are very different. Although the mother does not own her child (any more than the father does), and although mother/foetus bonding is by no means inevitable, nevertheless nine months of gestation can forge a far different relationship to and connection with the resulting baby than is created by two minutes of sperm 'donation'. It is not 'sexist' to suppose that there is a difference between the relationships of the biological father and the biological mother to their child, and for Baber to gloss over this difference is to misappropriate the concept of sexism.

Baber also claims not only that anyone can make a baby, but that most babies 'are conceived for their parents' sakes, to further their various ends'.[17] Even if this is true, however, it fails to justify 'surrogacy' arrangements. The fact that some people in other circumstances treat babies as means to ends is morally questionable, and not to be regarded with equanimity. It is part of the general tendency to treat children as products or artifacts, rather than as developing persons. It is, moreover, related to the process of commodification of babies to which Baber refers but which she never adequately confronts. Despite the attempt by many non-feminists to claim that the money in a 'surrogacy' arrangement pays for the right to rear a child, or for the reproductive services of the woman,[18] the fact remains that the full fee does not change hands unless and until a live child is surrendered. There is, therefore, a strong case to be made that in a 'surrogacy' arrangement, a baby is what is being bought and sold. In most other contexts, such an event would be correctly recognized as an instance of slavery. And no matter how well purchased infants may be treated, that fact does not justify baby-selling.

In addition, the fact that there are no 'intuitions and hormones' guaranteeing that biological mothers will act in their children's best interests[19] provides no reasons whatsoever for believing that 'surrogacy' is not detrimental to the babies produced. While there is no evidence to suggest, *prima facie*, that the babies would be better off with their biological mothers, there is also no evidence that they will be better off with the men who buy them. In fact, 'surrogacy' arrangements constitute a sort of ongoing psychological experiment on the women and children involved, and no one can predict its outcome. No one knows what the long-term consequences will be for the children produced in this way. Will they be particularly cherished and loved, because the parents had to go to such lengths to obtain them? Will they be subjected to unreasonably high parental demands and expectations, because the parents hope to receive good value for their money? So far, there are no answers to these questions, and Baber is at least highly misleading when she implies that no untoward consequences are to be feared.

Finally, Baber's reply to the second argument against 'surrogacy' is noteworthy for its elements of misogyny. She fails to give any recognition to women's reproductive labour, and sees nothing valuable in women's mothering. Worse still, she is convinced that women routinely seek to advance their 'self-interested ends'[20] through having children. She ignores the fact that pregnancy and mothering are compulsory for vast numbers of women, who lack any sort of choice about their reproductive future. And she grotesquely misplaces the power in most male-female relationships by describing women as 'bullying' reluctant partners.[21]

♦ ♦ ♦

Argument III

As Baber states this argument, which she calls 'The Practice of Surrogate Parenting Exploits Women', surrogate parenting arrangements are exploitative and thus against the interest of the women who act as surrogates. Surrogate mothers ... are compelled by their relative poverty and by their powerlessness as women in a sexist society to sell their services to richer, more powerful males.[22]

In response, Baber states that

exchange of goods or services is exploitative if and only if the following conditions are met:
(1) The situation of the seller must be one which would be recognized by him [sic!] and members of his community as, in some sense, abnormally bad.
(2) The price which the buyer pays must be substantially less than the value of the goods or services he procures.
(3) If the seller had not been in an abnormally bad situation he would not have sold his goods or services for the price offered by the buyer.[23]

Then, using the example of the Baby M case, Baber argues, first, that 'surrogate' mothers such as Mary Beth Whitehead are not in 'abnormally bad situations'. Second, she argues that the fee of $10,000 'for Mrs. Whitehead's services' 'was, if anything, generous', since Whitehead had no recent work experience and no skills; gestation is not skilled work, requires no special education or training, is 'not especially hazardous', and 'imposes far fewer constraints on a person's liberty than virtually any other job available'.[24] Finally, Baber reiterates that there is no evidence to suggest that all or most mothers 'bond' to their babies, or that those who give up their babies shortly after birth suffer extreme or long-lasting distress. She remarks:

there are many people who very much want to believe that women bond to their babies since many people have a vested interest in believing that women are very different from men psychologically and emotionally, that these differences are innate, and that they include a much greater stake in matter[s] related to reproduction and child rearing. Conservatives who wish to maintain traditional sex roles are included in this group but so are new style feminists who reject the old, liberal assimilationist ideal....[25]

With this argument, Baber explicitly claims to adopt a non-sexist perspective, while repudiating what she regards as the errors of some feminists. So-called 'new style feminists' are categorized with conservatives: both allegedly make the mistake of stressing women's differences from men and regarding those differences as inherent in women, and both allegedly are committed to the segregation of women from men, particularly in matters relating to procreation.

Since Baber never actually names any feminists or feminist works, it is difficult to know whom she has in mind. Her brief generalizations slide glibly over huge areas of debate within feminist theory. Nevertheless, in response to claims like these, it is important to reaffirm the value of one of the central strands of feminist thought: the emphasis upon women's own experiences, needs, beliefs, and wants. If, as seems plausible, these are different from men's, particularly with respect to reproduction, those differences were not invented by feminist analysis. What are the sources of those differences? Many feminists would cite both the social construction of sexuality and reproduction[26] and the biological differences (themselves subject to social construction) between women and men[27]—an explanation quite different from innatism.

In regard to the policy implications of these differences, feminists of course diverge radically from conservatives, since conservatives will advocate the maintenance of traditional women's roles, whereas feminists usually advocate that women recover the control over reproduction that men have been gradually arrogating to themselves.[28] In practice, the so-called 'assimilationist ideal' that Baber appears to cherish has usually resulted in the invisibility of women's experiences, belief systems, values, and activities and the predominance of the ways of 'man'.

Aside from this misrepresentation of feminist analysis, need we be persuaded by Baber's argument that women are not exploited in 'surrogacy' arrangements? Baber assumes once again, falsely, that there is nothing problematic in the conditions of work set out by the 'surrogacy' contract. She also assumes that what is purchased in 'surrogacy' arrangements is not a baby but the reproductive labour of the woman who is hired. She writes: 'Certainly, it seems morally repugnant to put any price on a human life. But the question is not whether Baby M was worth $10,000; the question is whether Mrs. Whitehead's services were worth that sum....'[29] As we have seen, the assumption that it is reproductive labour, not a baby, that is purchased in the 'surrogacy' contract is highly dubious, since a woman such as Mary Beth Whitehead who completes the labour of gestation but fails to surrender the product, the baby, does not get paid.

But even if we grant Baber's assumptions, for the sake of pursuing her argument, we need not accept her other claims. Baber does not define what she means by an 'abnormally bad situation', although she hints that the definition depends, in part, on the 'community's standards'.[30] In itself, this is a dangerous admission, since women have seldom participated in the formation of so-called 'community standards', and indeed often have little choice even about their membership in a given community. In addition, Baber is apparently oblivious to the class differences between the woman and the man who hires her.[31] There is already a fair amount of evidence to suggest that women who present themselves for 'surrogacy' arrangements are relatively disadvantaged as regards their education, income, and potential life opportunities, particularly when these are compared to those of the men who hire them.[32] Indeed, Baber herself grants that $10,000 is more than many of these women could otherwise earn for nine months' work.[33]

Yet Baber fails to ask herself whether producing a baby for hire is something these women would do if they had other opportunities for economic support and self-fulfilment. Is 'surrogacy' just a job like any other? Is it a choice we would want to promote for women? Would Baber herself (a university professor who, according to her one footnote, was seven months' pregnant with her third child at the time of writing[34]) be willing to sign a 'surrogacy' contract? If the answer to each of these questions is no, then Baber needs to rethink her claim that the situation of women who become 'surrogates' is not 'abnormally bad'. While feminists are committed to the development of a wide range of jobs and occupations and the provision of much greater choice for women among employment opportunities, they also point out that much of the work women have traditionally done is exploitive. To endorse women's 'choice' of 'surrogacy' because the work is not 'abnormally bad' is to appropriate the feminist commitment to job opportunity without acknowledging the feminist critique of exploitation.

While minimizing the disadvantages of women who enter 'surrogacy' contracts, and ignoring the class bias inherent in the arrangement, Baber shows extraordinary contempt for the work itself. This is, of course, consistent with her earlier undervaluing of maternal labour. To claim that $10,000 for reproductive labour is 'generous' is to endorse without question the patriarchal view of reproduction as merely a physical function, similar to excretion, without value or credit to the women who undertake it.

In addition, to state that Mary Beth Whitehead, already a mother of two children before she conceived Baby M, had 'no recent work experience and no skills'[35] is to say that mothering is not work and childcare requires no skills. Furthermore, Baber's claims that procreation is 'not especially hazardous' and that 'it imposes far fewer constraints on a person's liberty than virtually any other job available'[36] overlook the actual experience of pregnancy and childbirth in a way that is surprising, coming from a woman who has had these experiences. While it is true that many women do a great deal of other work while pregnant, and do not accept pregnancy as a limitation on their activities, nevertheless for many women pregnancy makes unique physical and emotional demands, and is a consuming experience that pervades their lives, waking and sleeping, for nine months.

Finally, it would be interesting to know on what evidence, if any, Baber founds her claim that those who surrender their babies suffer no serious or enduring distress. Because the practice is still so new, there are not yet any studies on the long-term feelings of hired mothers after surrendering their infants. Yet the accumulating anecdotal testimony of many such women attests to the pain of their loss.[37]

In conclusion, then, Baber's paper is a striking example of the use of pseudo-feminist language and arguments on behalf of a reproductive practice that oppresses women.[38] In exploring the many ethical and social-policy questions raised by reproductive engineering, it is important to be particularly vigilant about these pseudo-feminist views, and to recognize them for what they are: the co-optation of feminist discourse in the defence of patriarchal misuses of women's bodies.

Notes

[1] See, for example, Susan Crean, *In the Name of the Fathers: The Story Behind Child Custody* (Toronto: Amanita Enterprises, 1988).

[2] Baber, H[arriet] E. 'FOR the Legitimacy of Surrogate Contracts', in Herbert Richardson, ed., *On the Problem of Surrogate Parenthood: Analyzing the Baby M Case* (Lewiston, NY: Edwin Mellen Press, 1987).

[3] Ibid., 31.

[4] Throughout this chapter I have used the term 'surrogacy' to refer to the contract and the practice at issue, while trying to avoid the term 'surrogate mother', which, as many feminists have pointed out, is totally inappropriate for a woman who is a biological mother to her child. See, for example, Somer Brodribb, 'Delivering Babies: Contracts and Contradictions', in Christine Overall, ed., *The Future of Human Reproduction* (Toronto: Women's Press, 1989).

[5] Baber, 'Legitimacy', 32. 'Control of one's body' is, of course, a central value in the history of women's struggle for abortion. For discussion of this value in the American abortion rights movement, see Kristin Luker, *Abortion and the Politics of Motherhood* (Berkeley: University of California Press, 1984), and in the Canadian abortion rights movement, Anne Collins, *The Big Evasion: Abortion, the Issue That Won't Go Away* (Toronto: Lester and Orpen Dennys, 1985).

[6] However, it is not necessary for a supporter of a pro-choice position on abortion to be committed to the view that the foetus is merely a part of the mother's body. The classic statement of the argument that *even if a foetus is a person*, it has no right to the use of the pregnant woman's body, is found in Judith Jarvis Thomson's often-reprinted paper, 'A Defense of Abortion', *Philosophy and Public Affairs* 1 (Fall 1971): 47-66.

[7] Baber, 'Legitimacy', 33.

[8] Ibid., 31.

[9] For extensive criticisms of the exploitive nature of surrogacy contracts, see Gena Corea, *The Mother Machine: Reproductive Technologies from Artificial Insemination to Artificial Wombs* (New York: Harper and Row, 1985), 213-49.

[10] Baber, 'Legitimacy', 34.

[11]Ibid., 35.

[12]Ibid., 31.

[13]Phyllis Chesler appears to approach this view in *Sacred Bond: The Legacy of Baby M* (New York: Times Books, 1988).

[14]Christine Overall, *Ethics and Human Reproduction: A Feminist Analysis* (Boston: Allen and Unwin, 1987), 151-6.

[15]Mary O'Brien, *The Politics of Reproduction* (Boston: Routledge and Kegan Paul, 1981).

[16]'In the Matter of Baby "M", a Pseudonym for an Actual Person', 525 *Atlantic Reporter* (2d series, 1987), 1157.

[17]Baber, 'Legitimacy' 35.

[18]John A. Robertson is the most vociferous defender of this view. See his 'Procreative Liberty, Embryos, and Collaborative Reproduction: A Legal Perspective', in Elaine Hoffman Baruch, Amadeo F. D'Adamo, Jr, and Joni Seager, eds, *Embryos, Ethics, and Women's Rights: Exploring the New Reproductive Technologies* (New York: Haworth Press, 1988), 185-90.

[19]Baber, 'Legitimacy', 34.

[20]Ibid., 34.

[21]Ibid., 35.

[22]Ibid., 36.

[23]Ibid., 37.

[24]Ibid., 37-8.

[25]Ibid., 39.

[26]Amy Rossiter, *From Private to Public: A Feminist Exploration of Early Mothering* (Toronto: Women's Press, 1988).

[27]See Marilyn Frye, 'Sexism', in her *The Politics of Reality: Essays in Feminist Theory* (Trumansburg, NY: Crossing Press, 1983), 17-40, and Alison Jaggar, 'Human Biology in Feminist Theory: Sexual Equality Reconsidered', in Carol Gould, ed., *Beyond Domination: New Perspectives on Women and Philosophy* (Totowa, NJ: Rowman and Allanheld, 1983), 21-42.

[28]The feminist literature on the importance of women's reassertion of control over reproduction in response to the development of new reproductive technologies is vast. Two examples are Patricia Spallone and Deborah Lynn Steinberg, eds, *Made to Order: The Myth of Reproductive and Genetic Progress* (Oxford: Pergamon Press, 1987), and the anthology *Man-Made Women: How New Reproductive Technologies Affect Women* (Bloomington: Indiana University Press, 1987). For a recent and

thorough exploration of legal policy issues with respect to surrogacy, see Martha A. Field, *Surrogate Motherhood: The Legal and Human Issues* (Cambridge, MA: Harvard University Press, 1988).

[29]Baber, 'Legitimacy', 38.

[30]Ibid., 37.

[31]Sandra Anderson Garcia, 'The Baby M Case: A Class Struggle Over Undefined Rights, Unenforceable Responsibilities, and Inadequate Remedies', in Linda M. Whiteford and Marilyn L. Poland, eds, *New Approaches to Human Reproduction: Social and Ethical Dimensions* (Boulder, CO: Westview Press, 1989), 198-216.

[32]Daniela R. Roher, 'Surrogate Mothering' (unpublished manuscript, 1988).

[33]Baber, 'Legitimacy', 36.

[34]Ibid., 40, fn.

[35]Ibid., 37-8.

[36]Ibid., 38.

[37]Rita Arditti, 'A Summary of Some Recent Developments on Surrogacy in the United States', *Reproductive and Genetic Engineering: Journal of International Feminist Analysis* 1, 1 (1988): 60-4.

[38]Another recent example is Juliette Zipper and Selma Sevenhuijsen's 'Surrogacy: Feminist Notions of Motherhood Reconsidered', in Michelle Stanworth, ed., *Reproductive Technologies: Gender, Motherhood and Medicine* (Minneapolis: University of Minnesota Press, 1987), 118-38.

Chapter 7

♦ ♦ ♦

The Case Against Legalization of Contract Motherhood

'Surrogate motherhood' is a reproductive practice in which a woman agrees, before becoming pregnant, to surrender the baby she gestates to a man who has 'commissioned' the pregnancy, usually in return for a set fee. As many feminists have pointed out, the term 'surrogate motherhood' is inappropriate and ideologically misleading: it disguises and derogates the pregnant woman's reproductive labour. The woman who gestates the baby in these arrangements is a real mother to the resulting infant. In view of these observations, the practice of 'surrogacy' will in this chapter be referred to mainly as 'contract motherhood'.

The pregnancy is ordinarily the result of artificial insemination using sperm from the commissioning male, although it may also, less frequently, be the result of the implantation of a fertilized egg produced through *in vitro* fertilization, or recovered, through uterine lavage, from another woman's reproductive system. Where the pregnancy results from artificial insemination, the mother is a genetic parent of the baby. Where the pregnancy results from IVF or from uterine lavage, she is not. In the latter cases, the fertilized egg may be the product of gametes provided by the commissioning male and his wife or female partner, or in part or wholly of gametes donated by or purchased from other individuals.

In the United States, where most contract motherhood arrangements occur, the practice is usually facilitated by a third party or broker, who matches individual men or couples with women who are willing to act as their 'surrogates'. The broker oversees the selection of the women, their artificial insemination, their medical care during pregnancy and birth, and the contract specifying the rights and obligations of the parties involved. In return for those services, the commissioning party pays the broker a fee, which is usually substantial, in addition to the fee of the contract mother, and the costs incurred by her.

At least several hundred babies have been born through contract motherhood arrangements worldwide. Compared to the international and even the Canadian birthrates, these numbers are extremely small. One 1988 estimate placed the total number of cases of contract motherhood involving Canadians, whether as mothers or as commissioners of pregnancies, at 118.[1] Nevertheless, the practice itself is significant in terms both of its present implications for the status of women and children, and of its potential for much wider development, if it receives state sanction and support.

Defenders of contract motherhood often cite supposed biblical precedents for the practice. The book of Genesis tells the story of Sarah, who because of her infertility gave her 'handmaid', Hagar, to her husband, Abraham, to bear the child she herself was unable to have.[2] Aside from the dubious validity of using a religious argument to legitimize a secular practice, however, it is essential to note the disanalogies between the biblical case and modern arrangements. In the biblical case, Hagar presumably had no choice in the arrangements. Surely the proponents of contract motherhood do not want to force women to be breeders. Moreover, Hagar was impregnated through heterosexual intercourse (perhaps rape?) rather than through artificial insemination, as today, and she certainly was not paid even the meagre amount (usually $10,000) that today's contract mothers receive. Finally, on the most optimistic interpretation of the story, Hagar can perhaps be interpreted as agreeing, on altruistic grounds, to have a baby 'for' Sarah. Modern contract motherhood is very seldom an altruistic arrangement between two women. Instead, it is usually a commercial agreement, founded on a contract and arranged by a broker, in which a man hires a woman who is a stranger to him. The existence of biblical precedents in no way justifies present contract arrangements.

This chapter argues, from a feminist perspective, that Canadian social policy with respect to reproduction should *not* support contract motherhood; legislation should not be enacted to facilitate and regulate this reproductive practice. Indeed, there are good reasons for actively discouraging it, particularly in its commercial form.

The case against contract motherhood has already been argued quite extensively, if often haphazardly, by a variety of critics— from proponents of children's rights to conservative theologians— and not all are convincing. Some of the arguments are essentially anti-woman in nature. For example, it has been suggested that the legalization of contract motherhood would provide opportunities for feckless women to take advantage of desperate couples: '[W]ith increasing commercialization and a growing number of middle-class, politically and legally sophisticated women negotiating surrogate contracts, the potential for inequitable terms in these contracts seems substantial, as the ability to negotiate for top dollar and maximum rights takes precedence over any altruistic motives the women might have.'[3] As the discussion below will show, those who fear such opportunism overlook the inequitable power relationship between the contract mother and those who hire her, and underestimate the degree to which the woman can be coerced to abide by the wishes of those who pay her fee.

Other arguments against contract motherhood rely on religious or conservative moral assumptions about the family and sexuality. It is said, for example, that the practice violates the 'sanctity of the family', is 'unnatural', or is adultery. Some claim that contract motherhood violates the child's supposed right to be raised by the woman who gestated it. But there seems little reason to suppose that the nuclear family comprised of heterosexual parents and biologically related offspring is more 'natural' or 'sacred' than other forms; and such a supposition also easily works to the disadvantage of adopted children and of lesbian or homosexual parents and single parents.

Some feminists have cited the alleged bond between mother and infant as the determinative argument against contract motherhood: 'a child's own birth mother is meant for that child; ... premature physical separation from that mother—even by the child's genetic father—will cause trauma and injury that should be avoided.'[4] But not only does this claim by implication condemn all forms of adoption; it also explicitly relies on biological determinism—for example, on notions such as instinct and naturalness—in ways that are insufficiently justified by the data on mothering and that work against women's interests by condemning some to mothering in the alleged 'best interests' both of the children and of themselves. The case against contract motherhood must therefore rest on other grounds.

♦ ♦ ♦
Baby-Selling

Why should contract motherhood not be legalized? The first and most important reason is that contract motherhood as it is ordinarily practised involves the selling of babies. Defenders of 'surrogacy' usually deny this charge, variously arguing that couples actually are paying for 'salary for work done, ... compensation for the expenses involved in pregnancy, ... a gift intended to reciprocate the surrogate's generosity';[5] for parental rights to a child; for the 'willingness to be impregnated and carry [the] child to term';[6] or even for rental of the woman's womb.[7]

The term 'womb leasing' itself suggests that the mother's body is doing nothing; it is a mere container, the scene but not the means of reproduction. This impression is confirmed when it is claimed that '[a]t birth, the father does not purchase the child.... He cannot purchase what is already his.'[8] Yet in ordinary cases involving artificial insemination, if a sperm vendor showed up to claim 'his' baby, he would not be thought to own it or have any rights to it.[9] The fact that the man provided the sperm that resulted in conception does not make the baby straightforwardly 'his', and it does not invalidate the claim that the baby is being purchased.

Moreover, in the contracts themselves, it is neither rights nor opportunities that are being offered for sale, and the rights and opportunities to parenthood that are acquired by the new father cannot be exercised without possession of the child in question. In any case, parental rights are ordinarily not the sorts of things that can be bought or sold; they can only be severed, for good cause, by the state.[10]

That the product being sold is the baby itself is most strongly indicated by the fact that if a contract mother changes her mind and refuses to surrender the child, she may not be paid anything, let alone the full amount indicated in the contract. In addition, contracts usually provide that if the infant is stillborn, the mother will receive only $1,000. If it were in fact the use of the woman's uterus or her reproductive labour that was being purchased, then she should rightly be paid in full whether or not the baby survives the birth or is surrendered afterwards.

In almost any other context the exchange of money for possession of a human being would more easily be recognized for what it is: commerce in human slaves. Perhaps only because the human being in question is tiny, vulnerable, and unable to speak for itself are we unable to recognize a modern-day version of an old and grossly oppressive practice.

It is no answer to this argument to claim that the child will be loved and cared for. Perhaps this is true, although supporters of the argument have no empirical evidence to support it. Serving the best interests of the child that is 'produced' is in no way the goal of contract motherhood. The preconceptual agreement by the genetic parents of the child that one parent rather than the other will have custody of it violates ordinary procedures with respect to determination of custody, which is normally based on concern for the child. In fact, in so far as the best interests of the child are achieved through contract motherhood, that goal is at best a side effect; as will be argued later in this chapter, it is the best interests (as defined by him) of the hiring father—a wealthy and privileged individual—that are the overriding goal of the practice.

There is no reason to assume, because the child is the offspring of the man who provided the sperm and paid for it, that he will inevitably be a good parent for it. The 'best interests' of a child are not necessarily filled by the parent with the most money; there is no reason to believe that ability to pay defines the best parent(s) for a child. Merely wanting a child neither entitles one to raise a child nor enables one to provide and care for it well. On a speculative basis, it seems equally probable that a child who has been specially ordered and paid for by her parent(s) could be treated as an expensive commodity all her life—over-protected, perhaps, or subjected to unreasonably high expectations derived from her high purchase price. Such a child might one day be profoundly disturbed upon learning that she was bought and sold as an infant. Moreover, even if these speculations are false, and the purchased offspring of contract motherhood arrangements are all treated very well, the good treatment accorded slaves is never an adequate defence for the existence of slavery.

It has been argued that the 'slaves' in this case, though purchased, are 'freed' upon receipt, and acquire the rights and privileges accorded any other child in North American society:

> It is not really the case that if money passes in a surrogacy situation a baby becomes a thing. After all, in every other respect babies are subject to the laws governing people and not to the laws governing things or animals. And it seems wholly improbable that admitting commercial surrogacy would lead to slavery or anything resembling it.[11]

It has even been suggested that, since without the contract motherhood arrangement, these children would not exist at all, the arrangement confers on them a benefit beyond measure—or at least one that outweighs its drawbacks—even if they are born with disabilities caused by a drug-abusing mother![12]

Nevertheless, neither of these claims provides an adequate rationale for selling baby human beings. Such an arrangement amounts to a justification of baby farming: that the dubious morality of the process of generating the product is justified ultimately by the existence of the product itself.

But the argument here is not that selling babies leads, via the slippery slope, to slavery; the claim is that this practice *is* slavery, however well the babies may be treated. In Canadian society the sale of bodily organs, or even of that more plentiful commodity human blood, is neither legally permitted nor morally condoned. There is likewise no support for the 'farming' of human organs, or for enforcement of contracts to sell or even to donate an organ. If an impoverished person contracted to sell one of her bodily organs, no court would compel her to stand by the terms of the contract. Hence, if the sale of parts of human beings is not condoned, it is inconsistent to support the sale of a complete human being. The good that might be accomplished through the sale of human organs or blood or the marketing of human babies is not outweighed by the moral unacceptability of the commodification of body parts or of entire human beings.

◆ ◆ ◆

Reproductive Freedom and Exploitation

A frequent argument in favour of the legalization of contract motherhood involves an appeal to the need to protect the reproductive choice of both the buyers and the sellers. Access to contract motherhood is interpreted as essential to the alleged 'right to reproduce' of the infertile:

> Everyone ... has a right to a child; the infertile have the same rights as the fertile. Moreover, since artificial insemination is allowed so that a husband can procure a child legally his, it is unjust not to allow a wife to obtain a child which, as techniques improve, may even be genetically hers.[13]

Attorneys in the so-called 'Baby M' case made the following rights-based claims:

> The right of a married couple to engage in collaborative reproductive techniques such as surrogate parenting is constitutionally protected. It must be reasoned that if one has a right to procreate coitally, then one has the right to reproduce noncoitally. If it is the reproduction that is protected, then the means of reproduction are also to be protected.... Given the right to procreate is fundamental, governments cannot unduly interfere with the procreative liberty of participants to a surrogacy arrangement.[14]

Analogously, choosing to become a contract mother is interpreted as part of the 'reproductive freedom' of women: 'If a man may offer the means for procreation then a woman must equally be allowed to do so.'[15]

It is important, however, not to draw comparisons where they are inappropriate. In fact, the reproductive roles of males and females are not analogous, and sperm 'donation' is not remotely the same—in terms of time, energy, risk, or attachment—as pregnancy and birth.

Moreover, the reproductive-freedom argument is ambiguous: it fails to distinguish among different senses of 'right to reproduce'. As was noted in Chapter 1, the right to reproduce has two senses, weak and strong. The weak sense of the right to reproduce is the entitlement not to be interfered with in reproduction, or prevented from reproducing. It would imply an obligation on the state not to inhibit or limit reproductive liberty through forced sterilization, forced abortion, or coercive birth-control programs. But in its strong sense, the right to reproduce would be the entitlement to receive all necessary assistance to reproduce, including the gametes of other women and men, the gestational services of women, and the full range of procreative techniques, including *in vitro* fertilization, uterine lavage, and sex preselection.

The existence of a right not to reproduce and of a right to reproduce in the weak sense does not imply the existence of a right to reproduce in this stronger sense, especially if there are specific reasons why that right should not be recognized. In Canadian society, no right can require the work of women as breeders, or the subjugation of women's bodies to men or to the state. And the right to reproduce, in so far as there is one, is not the same as the right to custody and care of a child or children.

In any case, the availability of contract motherhood so far serves the reproductive freedom of only a very few. In defence of 'surrogacy' the sad plight of the infertile is often invoked: the longing for a baby, the empty arms, the years of vain efforts to have one's own offspring, the importance of 'forming a family'. And indeed, the miseries of the infertile are deep and undeniable. They are not, however, an unanalysable given. In the practice of contract motherhood, only the longings of the wealthy infertile seem to count. None of the defenders of this practice seem very concerned about the unhappiness of poor and working-class persons—who may, in fact, suffer from higher rates of infertility, often through greater exposure to reproductive hazards and contaminants. For the most part also, no one is very concerned about single persons, lesbian couples, people of colour, or disabled persons, who may 'want' a child just as much as do the wealthy, white, heterosexual couples who are able to buy a child through existing contract motherhood arrangements. Significantly, no one is arguing that these more marginal persons' access to contract mothers should be subsidized by the state via medicare plans. If hiring a contract mother is indubitably a valuable service, then its supporters should be prepared to say how it will be made available to all those who might want it.

Instead, 'surrogacy' caters to the baby-buying motives of only the very rich, who can 'afford' to pay up to $30,000 to the broker and the contract mother for this very special commodity. The point here is not that *ad hoc* limitations should be put on the spending patterns of the wealthy. Rather, it is that the reproductive freedom accruing to the rich through their ability to spend their money in this manner in no way constitutes sufficient justification for the practice—in particular, sufficient to compensate for its drawbacks.

In addition to understanding the feelings of those who are infertile, it is necessary to understand the nature and possible origins of those feelings. Feminists have recommended caution about accepting at face value the view that people have any 'innate' need to acquire children genetically related to them, or the view that this need, however acquired, must always be satisfied. The extensive ideology of pronatalism should encourage some scepticism about the existence of an inherent desire for children. In a society that teaches that one is not a real man or a real woman without a baby, that a child not genetically related to oneself is second best, and that people should go to any lengths to obtain their 'own' infant, it can be assumed that what appears to be a real 'need' on the part of infertile couples who hire contract mothers is at least in part a social artifact. Hence the desires of the wealthy infertile are not, in themselves, in any way a compelling and unanswerable argument for the justification of contract motherhood. Instead of encouraging and catering to the demand for contract mothers, social resources could be directed to exploring the causes and prevention of infertility, or to promoting other ways of relating to children.

Moreover, the desires that these arrangements cater to may be primarily those of the man in the hiring couple. The sperm provider, needless to say, does not want just any child; he wants one that is the product of his own sperm, and may even require that the newborn infant undergo tests to confirm that it is indeed his 'own'. (In light of this 'genetic narcissism',[16] there may be some cause for concern about the possible reactions of the sperm provider if the resulting child turns out to resemble its mother more than its father.) In order to avoid possible legal problems with respect to payment for adoption, the contract is usually an arrangement only between the man and the woman he hires. His wife, the presumably infertile person in this triangle, is a shadowy figure with no legal standing in the arrangement:

> There is a symbolic sexual replacement: she [the contract mother] is inseminated with the sperm of the other woman's husband. (She replaces the wife, and she abstains with her own normal partner, to avoid the risk of his being the biological father.)[17]

It cannot simply be assumed that the wife is as eager as her husband to acquire a child by this means. In some cases, she may not actually be infertile, or she may be infertile by choice; she may even already have children. Moreover, the wife's role as potential 'social mother' appears to be an acting out of female sex role stereotypes: 'A ... frequently posited characteristic of the social mother is her altruism: She wants her husband to be able to reproduce his own genetic material, even if she is not involved in the procreative process.'[18]

If little is known about the wife of the hiring father in such cases, however, a great deal of empirical information is available concerning the woman who is hired. These women are virtually always of a different class than the men who hire them. They often have little education, little or no income, and very little personal security. They are acting, in part, in response to the female socialization process that teaches them to be altruistic, generous, and compassionate as well as self-sacrificing, unassertive, and submissive to male authority.[19] They take the 'job' of contract mother out of a conviction that there is virtually nothing else that they can do:

> Surrogacy offers an economic alternative. It does not, however, prepare the surrogate mother for enhanced future job opportunities. Surrogate mothers are removed from the public job market during their pregnancy, thereby making it more difficult to find other employment after the pregnancy is completed.[20]

In other words, 'surrogacy' is an arrangement that manifests both class and sex inequalities.

Under these circumstances, where there are few, if any, perceived or real alternatives, it is appropriate to ask to what degree these women's choices to enter into contract motherhood are free ones. Some observers have questioned whether participants in contract motherhood are able to give informed consent to their involvement. 'It may not be possible to inform a surrogate of the full potential of psychological and emotional trauma she could experience when required to relinquish the child.'[21] Can a woman really know in advance what it will be like to surrender a child, this particular child, for money, particularly when she may know little or nothing about the person(s) to whom she must surrender it?

The contract itself typically exacerbates the exploitation these women experience. If interpreted as a fee for the woman's labour, the typical pay of $10,000 works out to far less than the minimum wage for a 24-hour-a-day 'job'. In fact, it is grotesquely disproportionate to the fee paid sperm 'donors', who usually receive $50 for two to five minutes of presumably not unpleasant 'work'. If the contract mother were likewise paid $10 per minute for her far more risky[22] reproductive labour, she would be entitled to a fee of at least $3,888,000.

The contract will usually specify limitations on the hired women's sexual activities, medical care, leisure time, travel, nutrition, and geographic location. It may require certain kinds of prenatal tests, and specify—in contravention both of law and of ordinary practice in both Canada and the United States—that the woman will abort or not abort at the behest of the hiring father. The peculiarity of this latter term of the contract is indicated by the fact that even an agreement by a wife with her husband not to abort their foetus is not enforceable.[23] If men are given contractual rights to claim the 'products of their sperm' by preventing the women they hire from aborting, there is good reason to be concerned about the precedent this might set for the limitation of other abortion rights and the expansion of so-called 'fathers' rights' (which could be more suitably dubbed 'impregnators' rights').

Finally, at the end of the pregnancy, in contrast to the provisions that govern ordinary adoption, the woman is not permitted to change her mind about surrendering the baby, an act that may cause her tremendous anguish. All of these extraordinary conditions raise the question of how compliance can be ensured: by means of imprisonment? restraint? forcible seizure of the infant?[24] For nine months, during which she assumes all the physical and psychological risks of pregnancy, the woman's body is not her own. This condition could easily be interpreted as a modern-day form of indentured servitude: the contract gives the sperm-provider hegemony over the contract mother's body. And, while it might be argued that a comparable power imbalance also occurs in other physically demanding jobs, only in 'surrogacy' is a person's body taken over by another, night and day, around the clock, for an unbroken nine-month period. One US observer has commented:

> If a 'surrogate' is *not* being paid for the baby, but only for her gestational 'services', then, according to state law, she is being grossly and illegally underpaid, i.e., she is not earning the minimum wage per hour, nor is she being paid in cash or on a weekly basis.
>
> If, on the other hand, a 'surrogate' *is* being paid to surrender the baby (a 'product'), then the contract violates both state and federal laws against baby-selling and against peonage or indentured servitude, i.e., a citizen and human being cannot be *forced* to perform against her will; nor can she be jailed for refusing to 'specifically perform'.[25]

Contracts for marriage, sexual services, or other 'personal services' are not specifically enforceable: that is, contract participants cannot be compelled to abide by the contract if they choose not to.[26]

Moreover, it is not possible to detach the handing over of the baby from the reproductive labour and argue that the former is enforceable even if the latter is not,[27] since this would still be baby-selling; it would not be intended for the child's best interests; and it would fail to allow any reconsideration period after the birth.

♦ ♦ ♦

Conclusion

Defenders of contract motherhood often argue that any ostensibly exploitive aspects of the contract can be mitigated through judicial means: for example, by increasing the hired woman's pay, by ensuring that the contract is genuinely protective of her interests, and by permitting her to 'change her mind' when it comes time to give up the baby. Others defend the funds paid to the working-class contract mother as 'financial resources that could easily purchase a major life-dream'.[28] The Ontario Law Reform Commission has recommended a number of lengthy legal provisions intended to legalize contract motherhood while avoiding some of its problematic aspects: for example, it recommends screening both the contracting couple and the prospective mother, and forbidding minors to be contract mothers.

However, merely mitigating the exploitive aspects of contract motherhood while making it legal begs the general question whether the practice as a whole is justified, whether the Canadian state should be fostering the work of women as breeders and whether this is a 'job' for women that Canadian society should endorse and support through state mechanisms. The legalization of contract motherhood would present reproduction for money as an acceptable, even desirable, aspect of women's place in Canadian society. But this path is incompatible with the vision of women as equal, autonomous, and valued members of this culture.

Do Canadians want the state to act as procurer of contract mothers? Do they want to entrench the use of women's reproductive labour to benefit male purchasers? Is this a direction for female employment that should be encouraged? Should girls be urged to consider contract motherhood as a possible career? (In fact, young women may be particularly vulnerable to exploitation in contract motherhood.) Does this practice promise a positive future for Canadian women?

Contract motherhood does not have a value that outweighs its negative effects. Aside from the message that the legalization of contract motherhood would convey both to women and about women, its effects on children also need to be considered. In a culture where children are still often treated as possessions, to be handled as private property and subjected to neglect, cruelty, and sexual assault, the legal entrenchment of contract motherhood would endorse the commodification of children, and legitimate the view that they are products to be disposed of at the discretion of their adult handlers. Some fear that legalization of and support for contract motherhood could have deleterious effects on the ordinary practice of adoption:[29] children who had not been specially ordered in advance and were not genetically related to the adoptive father could be regarded as less desirable. It is also important and appropriate to consider the short- and long-term effects on the siblings of babies produced through contract arrangements. It cannot be assumed that these children will not be affected by seeing the newest baby sold to another person. They may grieve over the loss of their little brother or sister, and their own personal security and feelings of belonging to the family could well be damaged.[30]

Finally, it is important to note the ways in which contract motherhood can serve to reinforce both racism and ableism. In contract motherhood arrangements, it is quite clearly not just any baby that is sought by the hiring father; it is a baby of a particular sort. When a man invests a lot of money in a product, he presumably wants one that is of high quality. He is likely to go to some trouble to choose a biological mother who represents the qualities he seeks in his offspring: perhaps an attractive appearance (where 'attractive' may have racist and sexist overtones), certain talents, and intelligence. This effort, of course, betrays a naïve over-confidence in the hereditary transmission of characteristics. Even more seriously, however, it betrays a eugenicist desire to obtain not just any baby, but the best possible one. Not only is a product ordered and purchased: it is a product of the very best sort. This demand is illustrated in a recent paper on contract motherhood, which cites the following supposed problems posed by the arrangement:

> Under what, if any, circumstances may the contracting couple refuse to accept the baby because of a birth defect? How long after the initial 'inspection' does the couple have to revoke acceptance if a defect, or 'nonconformity', is discovered?...What liability rests with the surrogate mother for genetic defects she passed on if she knew, or should have known, of the possibility of such an eventuality—for example, Down's syndrome, Tay-Sachs disease, sickle-cell anemia? What responsibility does the state bear to accept and care for babies who are rejected by parties to surrogate contracts?[31]

The explicit message here is that mentally and physically disabled children are not wanted. (Indeed, in the Stiver/Malahoff 'surrogacy' arrangement in the United States, a disabled baby was rejected by the hiring father.[32]) And even if the contract specifies that the purchaser must accept a child with a disability, the practice itself endorses the view of the perfectibility of children and the legitimacy of seeking only the very best in child-products.

Not surprisingly, in view of their economic characteristics, most purchasers of children through these contracts are white. They are not usually seeking to hire women of colour to produce their offspring. Given the oppressive nature of contract motherhood, this may well be a liability from which women of colour might be thankful to escape. But the combination of *in vitro* fertilization with contract motherhood creates more sinister possibilities, with clearly racist overtones. A white couple could arrange to have their own fertilized egg implanted in the uterus of a woman of colour who, because her opportunities are more limited, may be less expensive and more easily available than a white woman. The resulting baby would be a white child born of the reproductive labour of a woman of colour. It may thus be possible to drive down the cost of hiring a contract mother, and hence exacerbate the exploitation that women of colour already experience. Contract motherhood would provide a new way of exploiting the bodies of women of colour, while permitting white people to reinforce their racist preferences. The possible production

of a baby that is not genetically related to the woman who gestates it in no way obviates the major moral failings of the contract motherhood relationship, for once again there is real
exploitation of the woman who is hired, and money is paid for
the acquisition of a baby.

In the light of these arguments, it is clear that contract
motherhood should not be legalized. Supporters of legalization
often argue that failure to legalize will drive the practice 'under-
ground', creating even more opportunities for the exploitation of
women and affording no protections for the children produced.[33]
They claim that these practices simply cannot be stopped. The
assumption behind this argument, however, is that the practice is
likely to persist even when any state legitimation is withdrawn,
that there is some sort of inherent drive toward contract motherhood independent of the social climate in which it is created
and validated. But this claim conveniently overlooks the effects
of tolerance and legalization on the incidence of the practice.
Beyond isolated anecdotal reports of 'altruistic' surrogacy within
families, there is no evidence to indicate that contract motherhood arrangements, especially in their commercial form,
occurred with any great frequency before agencies and middlemen set out to create the market, recruit the women, and facilitate the buying of babies. But since the establishment of these
arrangements, the numbers have certainly been growing:

> [T]he publicity which has been accorded to surrogacy in the past
> few years has *legitimated* its use by people who might previously
> have considered it and then discounted it, or by those to whom
> the thought of asking another woman, whether within or without
> the family group, had not previously occurred.[34]

Although contract motherhood should not be legalized, the
individuals who seek out 'surrogacy' arrangements, especially
women who are recruited into them, should not be subjected to
criminal prosecution. But the work of the middlemen, the adver-
tisers, lawyers, doctors, and therapists who create and sustain the
industry, should be criminalized. Moreover, if some few 'surrogacy' arrangements are contracted, they should be rendered
unenforceable; at the end of the pregnancy, presumptive custody

would rest with the mother unless she is shown to be unfit or to have abandoned the child, since she is, literally, its prime care-taker. And the sperm provider would be liable for child support.

These steps would have the effect of discouraging contract motherhood by making such arrangements as unattractive as possible to all parties: there would be no certainty that the persons involved would get what they wanted. Contract motherhood may not disappear entirely, but refusing to legalize it would ensure that the numbers of cases are small.

Significantly, however, from a moral point of view, denying support to the 'surrogacy' industry does not in itself remove all that is objectionable about contract motherhood. Is so-called 'altruistic surrogacy' morally unproblematic? Recent cases include a seventeen-year-old girl/woman who gave birth to a child 'for' her mother, who was unable to conceive a child within her new marriage, and a 48-year-old grandmother who gave birth to triplets 'for' her daughter, who had undergone a hysterectomy after the birth of her first child. In these cases, the psychological coercion and exploitation of the contract mothers may arguably be wors-ened when they are not paid at all. And a baby is still being treated as an object, created and exchanged independently of its own needs and best interests, without the protective procedures associated with formal adoption. Furthermore, 'altruistic surroga-cy' leaves unchallenged the ideology of fertility, pronatalism, and emphasis on a genetic link with one's offspring and continues to foster a role for women as breeders. Hence, while the arguments presented in this chapter are primarily directed against commer-cial contract motherhood, there is no reason to be complacent about other forms of this practice.

Notes

[1] Dr Margrit Eichler, Ontario Institute for Studies in Education, letter, 16 Sept. 1988, p. 2.

[2] Genesis 16: 1-16.

[3] Sandra Anderson Garcia, 'Surrogate Mothering in the Marketplace: Will Sales Law Act as Surrogate for Surrogacy Law?' in Linda M. Whiteford and Marilyn L. Poland, eds, New Approaches to Human Reproduction: Social and Ethical Dimensions (Boulder, CO: Westview Press, 1989), 175.

[4]Phyllis Chesler, *Sacred Bond: The Legacy of Baby M* (New York: Vintage Books, 1989), 144-5.

[5]Jonathan Glover et al., *Ethics of New Reproductive Technologies: The Glover Report to the European Commission* (DeKalb: Northern Illinois University Press, 1989), 69.

[6]'In the Matter of Baby "M", a Pseudonym for an Actual Person', Superior Court of New Jersey, 525 *Atlantic Reporter* 2d Series (1987): 1157.

[7]'Nothing Left to Chance in "Rent-a-Womb" Agreements', *Toronto Star*, 13 January 1985.

[8]Ibid.

[9]Barbara Katz Rothman, *Recreating Motherhood: Ideology and Technology in a Patriarchal Society* (New York: W.W. Norton, 1989), 232-3.

[10]Chesler, *Sacred Bond*, 114.

[11]Sybil Wolfram, 'Surrogacy in the United Kingdom', in Whiteford and Poland, eds, *New Approaches*, 193.

[12]Glover, *Ethics*, 74, 75; 'Heavy smokers, alcoholics and other addicts may harm the child and these conditions should be grounds for exclusion [from eligibility to be a contract mother]....Where surrogates are very hard to obtain, these grounds *may* perhaps be overridden. From the later perspective of the child, it may be better to have run those risks, or even to have suffered some harm, than not to have been born at all' (Glover, p. 81, his emphasis).

[13]Wolfram, 'Surrogacy', 191.

[14]Sandra Anderson Garcia, 'The Baby M Case: A Class Struggle Over Undefined Rights, Unenforceable Responsibilities, and Inadequate Remedies', in Whiteford and Poland, eds, *New Approaches*, 214.

[15]'In the Matter of Baby "M"', 1165.

[16]Chesler, *Sacred Bond*, 20.

[17]Glover, *Ethics*, 67.

[18]Linda M. Whiteford, 'Commercial Surrogacy: Social Issues Behind the Controversy', in Whiteford and Poland, eds, *New Approaches*, 156.

[19]Elizabeth Kane, 'Surrogate Parenting: A Division of Families, Not a Creation', *Reproductive and Genetic Engineering: Journal of International Feminist Analysis* 2, 2 (1989): 105-6.

[20]Whiteford, 'Commercial Surrogacy', 152.

[21]Ibid., 153.

[22]These risks include the possibility of acquiring a sexually transmitted disease or even Acquired Immunodeficiency Syndrome (AIDS).

[23]Martha A. Field, *Surrogate Motherhood: The Legal and Human Issues* (Cambridge, MA: Harvard University Press, 1988), 65.

[24]Garcia, 'The Baby M Case', 205.

[25]Chesler, *Sacred Bond*, 113, her emphasis.

[26]Field, *Surrogate Motherhood*, 79.

[27]Lori B. Andrews, 'Alternative Modes of Reproduction', in Sherrill Cohen and Nadine Taub, *Reproductive Laws for the 1990s* (Clifton, NJ: Humana, 1989), 385.

[28]Michael R. Hill, 'A Cross Cultural Analysis of Several Forms of Parenting: Mother, Genetrix, and Mater', in Herbert Richardson, ed., *On the Problem of Surrogate Parenthood: Analyzing the Baby M Case* (Queenston, Ont.: Edwin Mellen, 1987), 80.

[29]Field, *Surrogate Motherhood*, 57.

[30]Whiteford, 'Commercial Surrogacy', 154-5.

[31]Garcia, 'Surrogate Mothering In the Marketplace', 180.

[32]See Case 3.2 in 'Cases For Preliminary Discussion', in Richard T. Hull, ed., *Ethical Issues in the New Reproductive Technologies* (Belmont, CA: Wadsworth, 1990), 150.

[33]Ontario Law Reform Commission, *Report on Human Artificial Reproduction and Related Matters* II (Ontario Ministry of the Attorney General, 1985), 232.

[34]Derek Morgan, 'Surrogacy: An Introductory Essay', in Robert Lee and Derek Morgan, eds, *Birthrights: Law and Ethics at the Beginnings of Life* (London: Routledge, 1989), 71, his emphasis.

Chapter 8

◆ ◆ ◆

Access to *In Vitro* Fertilization:
Costs, Care, and Consent

What would be a genuinely caring approach to the provision of
procedures of so-called 'artificial reproduction' such as *in vitro* fertilization? What social policies are appropriate and justified with
respect to attempting to enable infertile persons to have offspring?
These urgent questions have provoked significant disagreements
among theologians, sociologists, health-care providers, philosophers and even—or especially—among feminists. In the existing
literature and in developing social policy, three different kinds of
answers can be discerned: (1) some have suggested that access to
IVF should be provided as a matter of right; (2) some existing
social policies and practices imply that access to IVF is a privilege;
and (3) some theorists have argued that, because of its alleged violation of family values and marital security, or because of its risks,
costs, and low success rate, IVF should not be available at all. After
evaluating each of these views, this chapter will offer a feminist
alternative, describing what I think would constitute the caring
provision of *in vitro* fertilization.

◆ ◆ ◆
Access to Artificial Reproduction as a Right

Is there a moral *right* of access to *in vitro* fertilization? Some
feminists are remarkably suspicious of any use of rights talk by a

feminist, particularly in the context of reproductive technology.[1] While talk of rights does not exhaust feminist moral and political discourse about reproduction, and while appeals to rights can sometimes be used against women (for example, the appeal to the supposed 'right' to be a 'surrogate' or contract mother), surely the history of feminist activism with respect to abortion provides some indication that rights claims may still be useful, and that claims about reproductive rights need clarification.

As in the case of contract motherhood (Chapter 7), it is essential to distinguish between the right to reproduce and the right *not* to reproduce.[2] The latter means the entitlement not to be compelled to beget or bear children against one's will; to donate gametes or embryos against one's will; or to engage in forced reproductive labour. This right mandates access to contraception and abortion.

In its weak sense, the right to reproduce is the negative or liberty right not to be interfered with in reproduction, or prevented from reproducing. Thus it would obligate the state not to inhibit or limit reproductive liberty through, for example, racist marriage laws, fornication laws,[3] or forced sterilization or abortion.

In its strong sense, however, the right to reproduce as a positive or 'welfare' right would imply entitlement of access to any and all available forms of reproductive assistance, including the reproductive products and gestational services of other women— an entitlement that could violate the latter's rights *not* to reproduce. For this reason (examined in detail in Chapter 1), I conclude that there is no right to reproduce in the strong sense, and hence no justification for treating access to methods of artificial reproduction, such as IVF, as a right.

◆ ◆ ◆

Access to Artificial Reproduction as a Privilege

The absence of a right in the strong sense to IVF does not of course imply that all use of IVF is therefore unjustified. However, if we abandon the appeal to such a right, it may seem that we are committed to the view that having children by means of artificial reproduction is necessarily a *privilege* that must be earned through the possession of certain personal, social, sexual, and/or financial characteristics. The provision of reproductive technology then appears to become a luxury service, access to which can be controlled by means of the criteria used to screen potential candidates.[4] Such limitations appear to be the price of sacrificing a right to reproduce in the strong sense.

And indeed, in actual practice, the criteria of eligibility for processes such as IVF have included such characteristics as sexual orientation (only heterosexuals need apply); marital status (single women are not usually eligible, unless they are part of an ongoing marriage-like relationship[5]); and consent of the spouse. Because IVF is costly, economic status and geographical location have also become, at least indirectly, criteria of eligibility. We can speculate that these are likely to lead to *de facto* discrimination against working-class women and women of colour.[6] Further criteria have also been used—for example, reproductive age, absence of physical disabilities, and characteristics such as 'stability' and parenting capacities.[7] Some have also suggested or implied that infertility that is the result of the patient's own choice (for example, tubal ligation) should render the patient ineligible for IVF.

Should access to IVF be treated as a matter of privilege rather than right? Three arguments tell against this approach. First, persons who do not have fertility problems are not compelled to undergo any evaluation of their eligibility for parenthood. Moreover, some medical responses to infertility—for example, the surgical repair of damaged fallopian tubes—are undertaken without any inquiry into the patient's marital status, sexual orientation, or fitness for parenthood. If *in vitro* fertilization is classed as a medical procedure in the way that tubal repair is a medical procedure, then discrimination in access for the former and not for the latter is unjustified. The case of IVF seems to present an instance of discrimination on the basis of social criteria against people with infertility—and only certain kinds of infertility at that.

A second argument is the general difficulty of assessing the presence of some of the characteristics that have been assumed to be relevant for access to IVF. For example, for some women sexual orientation is a fluid and changing personal characteristic.[8] In addition, it is difficult to see how 'stability' or aptitude for parenthood can be adequately measured, and there is likely to be a lot of disagreement about the appropriateness of criteria for evaluating these characteristics. One could also challenge the justification for entrusting the assessment of these characteristics to IVF clinicians, who are not likely to be any better equipped than the rest of the population to make such evaluations.

Finally, it is essential to challenge the moral legitimacy of discrimination on the basis of characteristics such as sexual orientation and marital status. Such discrimination is founded upon false assumptions about the nature and abilities of single and lesbian women, and about the kind of mothering they can provide. While promoting good parenting practices is indisputably a worthwhile social goal, there is no evidence to suggest either that marriage and heterosexuality necessarily make women better mothers, or that the presence of a father is indispensable to childhood developmental processes. Nor do any research findings indicate that the ability to pay the enormous financial costs of IVF increases one's capacity to be a good parent.

Thus many purveyors of IVF (at least in Canada) seem to be guilty of an inconsistency. Medical practice is not usually premised on the assumption that only some patients deserve treatment, and IVF clinicians themselves see their role as one of relieving a disability or responding to an 'illness' in infertile women.[9] Yet they are willing to treat only those infertile women with social characteristics that they judge acceptable, and they disregard the experiences and needs of other infertile women who fail to conform to their criteria. There is no adequate justification for making access to procedures such as *in vitro* fertilization a privilege for which it is legitimate to erect social barriers that discriminate on arbitrary and unfair grounds—grounds such as marital status, sexual orientation, putative stability or parenting potential, or economic level. Moreover, given that IVF is still, as I shall argue below, an experimental procedure rather than an established medical practice, it is particularly unjust to exact money from those women whose bodies function as experimental material.

Accepting these social barriers to accessibility is not the only alternative to claiming a right of access to IVF. Rather, we should critically evaluate screening processes for IVF, and resist and reject practices of unjustified discrimination in access.

♦ ♦ ♦

Calling a Halt to Artificial Reproduction

Conservative Religious Arguments

Some critics of artificial reproduction regard access as neither a right nor a privilege; instead they condemn research in IVF and call for an end to such services. There are two very different reasons for this perspective.

On the one hand, some writers, particularly those influenced by the teachings of the Roman Catholic Church,[10] claim that *in vitro* fertilization threatens marital relationships, sexual interactions, and the integrity of the nuclear family. One representative of this approach is Canadian philosopher Donald DeMarco, who states:

> IVF demands sundering flesh from spirit in an area where the integrity of parenthood demands they be one, and sundering [sic; perhaps 'surrendering'?] that flesh to the manipulation of technicians. Inevitably, something important, though unseen, stands to be harmed in the process. And what stands to be harmed is human parenthood.[11]

As a mother myself, I have seldom found that human parenthood is 'unseen'. However, DeMarco explains further:

> By removing the child from the personal context of conjugal love, as IVF does, a decisive step is taken which necessarily depreciates that love.... And to weaken this love which is the essential bonding act of the family ... is to weaken the family. And since the family is the basic unit of society, what weakens the family also weakens the society.[12]

This set of claims is highly implausible. There is no evidence of an appreciable debilitation of the nuclear family attributable to the use of IVF. If anything, as many feminists have pointed out, the use of IVF strengthens the traditional nuclear family,[13] since it is usually provided only or primarily to persons who are part of heterosexual married couples, and it does not challenge the traditional belief that a family is not a real family without one or more genetically related children. Moreover, the legitimacy and value of adoptive relationships is implicitly and unjustifiably called into question by DeMarco's argument, since adopted children are not linked to their social parents through 'the personal context of conjugal love'. Finally, there is in DeMarco's claims a peculiar emphasis on the physical/material aspect of married heterosexual relationships, an emphasis that takes on an especially sinister aspect when DeMarco assures his readers elsewhere that '[h]usband and wife do have a *right* to engage in intercourse with each other',[14] suggesting that he does not recognize marital rape. DeMarco's belief that IVF 'degrad[es] the two-in-one flesh unity of parents by deflating the importance of the flesh as a vehicle of love in the formation of new life'[15] suggests that heterosexual intercourse has an extraordinary vulnerability to technological degradation, one that most of us would never have imagined. After all, nothing in the provision of IVF prevents heterosexual married couples from continuing to have sexual intercourse. DeMarco's claims implicitly condemn any intercourse (such as that involving contraception, or in cases where one partner is not fertile) that lacks the potential to result in conception. For all these reasons, DeMarco's reservations about IVF are not persuasive, and the call by religious conservatives for a ban on IVF lacks justification.

Feminist Arguments

At the other end of the spectrum of general opposition to IVF are criticisms expressed by some feminist scholars, scientists, and activists, criticisms that carry considerable empirical weight. For example, Canadian journalist Ann Pappert has investigated the sorry success record—perhaps more appropriately called a failure record—of IVF in Canada and the United States. She states: 'Of the more than 150 IVF clinics in the United States, half have never had a birth, and only a handful have recorded more than five. Fifty per cent of all U.S. IVF babies come from three clinics.'[16] The success rate at the best IVF clinic in Canada is 13 per cent; the majority of Canada's twelve IVF clinics have success rates of 8 per cent or lower.[17]

The stressful and debilitating nature of the IVF experience for women has been powerfully documented by Canadian sociologist

Linda Williams:[18] among the psychological costs are depression, anxiety, and low self-esteem. But the costs in physical suffering and consequent health care are even worse. They include the adverse effects of hormones such as Clomid, which are usually taken in large, concentrated doses to stimulate hyperovulation; repeated anesthesia and surgery to extract eggs; the heightened risk of ectopic pregnancy (when the fertilized egg implants outside the uterus); the development of ovarian cysts and menstrual difficulties; and the early onset of menopause and an increased risk of some forms of cancer.

Moreover, while it is often said that the children produced through IVF are healthy, some recent studies in Australia dispute that claim. Rates of multiple pregnancy, spontaneous abortion, pre-term delivery, perinatal death, birth defects, and low birthweight are higher in IVF than other pregnancies.[19]

Because IVF represents an ongoing medical experiment on women and children—an experiment whose first success, Louise Joy Brown (born in 1978), is still an adolescent—its long-term effects and risks are not known. Anita Direcks, a 'DES daughter' from Holland, has written movingly about the parallels between the use of the synthetic hormone diethylstilbestrol (DES) (intended to prevent miscarriage) during the nineteen forties, fifties, and sixties, and the use of *in vitro* fertilization during the seventies and eighties. Direcks writes:

> IVF is delivered by the same men who brought us DES, dangerous contraceptives, and other fertility-destroying technologies. One of the most important concerns I have in regard to IVF is the concern about the long-term effects of an IVF treatment for mother and child: the consequences of the hormonal treatment, the medium, and so on.... IVF is an experiment on healthy women.[20]

Indeed, the parallels between the development and use of DES and the development and use of IVF are alarming. DES was not adequately tested before being used on thousands of women; IVF was not adequately tested (not even on animals) before being used on thousands of women.[21] The long-term effects of DES were not widely known or were ignored when the drug was first being prescribed; the long-term effects of IVF are still not known. DES has intergenerational effects; there is a possibility that IVF may have intergenerational effects, especially in view of the extensive use of hormones in the generation of test-tube babies. DES was not effective in preventing miscarriages, but this was not made known to the public; IVF has a very low success rate, but this is almost systematically hidden from public awareness. Women using DES were not adequately informed about it; women undergoing IVF are not adequately informed about it. DES was recommended for routine use in all pregnancies, supposedly to produce 'better babies'; similarly, some of the promoters and defenders of IVF have claimed that the process can be used to 'reduce the incidence of, or eliminate, certain defects from the population'.[22] Thus the potential uses for DES were gradually and needlessly expanded, just as the potential uses for IVF are being gradually, and perhaps needlessly, expanded to include use in cases of male infertility where donor insemination would work just as well.

Second, there are significant similarities in the ideological underpinnings of the development of DES and IVF. These include the idea of the inadequacy of women's bodies; the goal of improving women's reproductive functioning; the emphasis upon science and scientists as the white knights coming to rescue women from their underfunctioning reproductive systems; the accent upon doing everything possible in the attempt to produce a baby, genetically related to oneself; the eugenic emphasis on having the perfect baby; and the ongoing focus on fertility and reproduction as central to, and perhaps definitive of, women and womanhood.

As a result of considerations such as these, some feminists have called for a ban on further IVF research and practice. For example, Renate Klein and Robyn Rowland state that 'IVF—in all its forms—must be ... abandoned. It is a failed and dangerous technology. And it produces a vulnerable population of women on which to continue experimentation.'[23] FINRRAGE, the Feminist International Network of Resistance to Reproductive and Genetic Engineering, calls for resistance to 'the development and application of genetic and reproductive engineering' and to 'the take-over of our bodies for male use, for profit making, population control, medical experimentation and misogynous science'.[24]

Feminists who would ban IVF see those women who use it in an entirely different light than do the liberals who identify IVF access as a right. In the view of the former, far from being free and equal contractors in the reproductive marketplace, women are victims, the uncomprehending dupes of the scientific and medical systems. Whereas the rights advocates regard IVF as inevitably serving women's reproductive autonomy, advocates of a ban regard IVF as inevitably destroying it. Whereas the rights advocates claim that women want IVF, advocates of a ban claim that women do not (really) want it, or that their desire for IVF is artificially created:

> Does not this obsessive craving to have a child of one's own in many cases stem from an individual's sense of private property or the desire to have somebody around over whom one has substantial control for some years at least? Let us also face the questions that (a) is not this craving more created than natural and (b) does not the social pressure to fit ... the image of 'motherhood' put women in a more vulnerable position?[25]

But while many feminists have rightly stressed both the social construction of the desire for motherhood and the dangers and ineffectiveness of *in vitro* fertilization,[26] not all of them have been willing to attribute women's desire for IVF simply to false consciousness. As Margarete Sandelowski writes:

> Feminists critical of the new conceptive technology and certain surrogacy and adoption arrangements suggest misguided volition on the part of infertile women, a failure of will associated not with causing infertility but with seeking solutions for it deemed hazardous to other women.... Beyond being politically useful as evidence for women's oppressive socialization to become mothers and their continued subservience to institutionalized medicine, infertile women occupy no more empathic place in many current feminist discussions than in the medical and ethical debates on reproductive technology feminists criticize.[27]

Sandelowski argues that some feminist theorists 'equate women's desire for children with their oppression as women, viewing this desire and the anguish women feel when it remains unfulfilled as socially constructed rather than authentically experienced'.[28] Thus women's desires are discounted and their autonomy is denied through the designation of socialization as the shaper and moulder of female selves.

Similarly, Christine St. Peters argues:

> The appeal to resist [the social imperative that women achieve personhood only through motherhood], an appeal that is heavily pedagogic in tone, is a staple of virtually all the feminist discussions of female infertility, which generally argue that the desire for motherhood is socially constructed and therefore susceptible to revision. Of course this is demonstrably true, although to what extent we cannot prove, since we cannot definitively demonstrate where nature and culture are separable. But the limitations of the message are particularly obvious at a strategic level where we must respond to infertile women's suffering; here the often homiletic tone probably alienates many women, especially as we have not yet changed the social contexts in which the desire for children takes the shape of desiring genetically related offspring.[29]

In fact, women's motives for seeking IVF are complex,[30] and it is important not to deny or underestimate the needs and experiences of infertile women.[31] It is surely inappropriate for feminists to claim to understand better than infertile women themselves the origins and significance of their desire for children. Even if the longing felt by infertile women is socially produced, it is nevertheless real. Furthermore, that longing cannot be assumed to extinguish women's autonomy. Women who will 'try everything' in order to obtain a baby are not necessarily less autonomous, less free from social conditioning, than women who gestate and deliver without technological intervention—or than the feminists who call into question infertile women's motivations.

Sociologist Judith Lorber claims that consent to IVF is not a freely chosen act unless the woman is 'an equal or dominant in the situation'.[32] But if that is the criterion for freedom of choice, then almost no women make free choices, ever. Philosopher Mary Anne Warren is more persuasive when she claims: 'Freedom is not an all or nothing affair. We can rarely be completely free of unjust or inappropriate social and economic pressures, but we can sometimes make sound and appropriate decisions, in the light of our own circumstances.'[33]

Radical feminist Janice G. Raymond has poured scorn on the kind of approach I advocate here, which she dismissively labels the 'nuanced' approach to evaluating reproductive technologies.[34] According to Raymond, advocates of this approach seek to 'limit the abuse [of women by reproductive technologies] by gaining control of some of these technologies, and by ensuring equal access for all women who need/desire them.' Their error, she suggests, lies in conflating need with desire, and then claiming that to oppose such needs/desires is to 'limit women's reproductive liberty, options, and choices'; in her view, 'women as a class have a stake in reclaiming the female body—not as female nature—and not just by taking the body seriously—but by refusing to yield control of it to men, to the fetus, to the State....'[35]

In fact, I believe that as feminists we can be extremely critical of the easy equation of need and desire, and of the social processes that, having created women's alleged 'need' for babies, require that that 'need' be fulfilled through a biologically related infant acquired in any way possible. We can also reject the facile claim that access to any and all reproductive services, products, and labour is indispensable to reproductive freedom. But it does not follow that feminists should protect women from these social processes and from acting on their own desires. We need not take women's desires as an unanalysable and unrejectable given. But neither can we ignore or belittle what women say they feel. We can attack the manipulation of women's desires by current medical/scientific reproductive practices. But we can also resist the too-simple depiction of infertile women as mere dupes or victims.

Raymond claims that radical feminists who expose the victimization of women by men are inappropriately 'blamed for creating' that victimization.[36] Of course feminists did not create the harm of IVF to which they have called attention. What I am suggesting is that feminists can expose the harmful aspects of IVF to the women who are most likely to be affected by it, and then let them decide themselves whether to seek access nevertheless.

The demand for an end to all use of IVF is an expression of a kind of feminist maternalism[37] that seeks to protect the best interests of the women affected by IVF. I cannot agree with those who wish to ban IVF to protect women from the dangers of coercive IVF, any more than I can agree with so-called 'pro-life feminists' who wish to ban abortion to protect women from the dangers of coercive abortions. It is not the role of feminist research and action to protect women from what is interpreted to be their own false consciousness. If, as Judith Lorber claims, women seeking IVF make 'a patriarchal bargain' rather than a free choice,[38] then those women must be given the information and support they need in order to genuinely choose.

At present, although women candidates are told something about the mechanical procedures for IVF, there is not much evidence that they are fully informed about the low success rates and the suffering and risks associated with the process. In such cases, the solution to ideological coercion is not necessarily to ban the option in question, but to educate people about what they are choosing.

Therefore, while I cannot support and endorse highly ineffective, costly, and painful procedures such as IVF, until infertile women themselves by the thousands, and especially those who seek and have sought IVF, call for the banning of artificial reproduction I cannot endorse such a call by some feminists any more than I would support a ban on all interventionist hospital births in low-risk deliveries. I assume that when women are provided with complete information, real choices, and full support with regard to artificial reproduction, they will be empowered to make reproductive decisions that will genuinely benefit themselves and their children. My subjective impressions, gained from talking to women who have either rejected IVF or have tried it and now have serious criticisms, suggest that, when fully informed, women may well reject *in vitro* fertilization at a much higher rate than they do now.

◆ ◆ ◆

A Feminist Alternative

On the other hand, they may not. It is therefore necessary to consider what might constitute a caring feminist approach to the provision of *in vitro* fertilization—an approach founded upon women's experiences, values and beliefs, that acknowledges the political elements of reproductive choices and practices, that seeks to minimize harm to women and children, and that recognizes and fosters women's dignity and self-determination. The caring provision of artificial reproduction services requires (a) truly informed choice and consent; (b) equal and fair access, unbiased by geographic, economic, or social criteria; (c) adequate record-keeping, follow-up, and research; and (d) appropriate support systems for all participants. All of these services could be provided in free-standing women's reproductive health clinics, run on feminist principles, where the health-care providers are both responsible and responsive primarily to their women clients.[39]

First, then, it is necessary to ensure that every woman—as an individual, not as part of a couple—entering and participating in an infertility treatment program has made a genuinely informed choice. Counselling should be provided not by the clinic itself, but by third parties who have no personal investment in persuading clients to use the clinic's services.

The notion of informed choice involves not merely telling women of the possible risks of the procedure, but discussion of the alternatives to *in vitro* fertilization.[40] It would also require open acknowledgement of the experimental status of the procedure. This point has recently been emphasized by Marsden Wagner, Director of Maternal and Child Health, European Regional Office of the World Health Organization: 'There has not been one single prospective, randomized controlled trial of the efficacy and safety of [IVF].... IVF is clearly an experimental procedure by all criteria. It should not be included in the health *care* budget, but in the health *research* budget.'[41]

Prospective patients must therefore be informed of IVF's unknowns, the short- and long-term risks, the possible benefits, the chances of success and failure, alternative approaches and treatments, and pronatalist social pressures to procreate and other ways of responding to them. In particular, women who are offered *in vitro* fertilization because their male partners are infertile should clearly understand that they could become pregnant much more easily, and with lower risks, if they made use of donor insemination.[42]

Second, it is essential to critically examine the artificial criteria, such as marital status, sexual orientation, and ability to pay, that get in the way of women's fair access to reproductive technologies, with a view to dismantling those barriers that discriminate unjustifiably. *If* IVF is a valuable medical service (and, given its high risks and low success rate, that assumption will remain debatable) then it deserves to be made available, like other medical services, through medicare, as it is now in Ontario.

Third, an adequate system of record-keeping should be established, to track the long-term effects of IVF on women and their offspring, and to ensure that any women who provide eggs for the program have genuinely chosen to do so, in order to eliminate 'egg-snatching'. Moreover, donors should really be donors, not vendors; the commodification of reproductive products and services is morally unjustified. It is essential to resist the commercialization of reproduction and the spread of reproductive entrepreneurialism, the primary targets of which are likely to be poor women and women of colour. It would also be important to ensure thorough screening and long-term follow-up of donors of eggs and sperm, and to avoid too-frequent use of the same donors. The issue of control over and decision-making about so-called 'spare' embryos, including those that are subject to cryopreservation, must be also be faced. In addition, offspring of artificial reproduction need certain protections: in particular, access to information both about their origins and the health status of their biological parents (if they were conceived using donor gametes), and about the implications of IVF—many of which are at present unknown—for their own health throughout their lives.

Finally, participants and potential participants in IVF programs should be provided with support systems that will enable them to fully evaluate their own reasons and goals, and provide assistance throughout the emotionally and physically demanding aspects of the treatment. Counselling and group support should function not merely as a means of ensuring the patients' continued acquiescence, or eliminating those without the stamina to endure the ordeal,[43] but as a means of facilitating their active involvement and participation in their treatment.

The approach sketched here is a fair and caring approach to the justification of and access to *in vitro* fertilization. It avoids, on the one hand, claiming access to artificial reproduction as a right in the strong sense, and on the other hand, making access to reproductive technology a privilege to be earned through the possession of certain personal, social, sexual, and/or financial characteristics. Sweeping generalizations about the moral justification of all forms of artificial reproduction are on very uncertain ground: processes of artificial reproduction need to be evaluated individually, on their own merits, to determine which ones, if any, are genuinely valuable and worth supporting.

This discussion has set aside some crucial questions of allocation and the relative importance of IVF in comparison with other health-care services, particularly infertility prevention, prenatal care, research on AIDS, and sex education. I do not assume that IVF inevitably ranks equal in importance with these other measures, or that IVF could not legitimately be limited—perhaps by confining it to those who do not have any children already, or by eliminating IVF as a so-called 'treatment' for male infertility —for the sake of research in and access to other, more pressing health services. Over the long term, certainly, the caring provision of artificial reproduction should be coupled with research into the incidence of infertility; into elimination, where possible, of its causes (e.g., previous medical interventions and environmental hazards); and into cures, so that the apparent need for artificial reproduction is reduced. Ultimately, the genuinely caring provision of artificial reproduction will require a feminist re-evaluation and reconstruction of all reproductive values, technologies, and practices.

Notes

[1]For example, see Janice G. Raymond, 'Reproductive Technologies, Radical Feminism, and Socialist Liberalism', *Reproductive and Genetic Engineering: Journal of International Feminist Analysis* 2, 2 (1989): 141.

[2]See also Christine Overall, *Ethics and Human Reproduction: A Feminist Analysis* (Boston: Allen and Unwin, 1987), 166-96.

[3]Ethics Committee of the American Fertility Society, 'The Constitutional Aspects of Procreative Liberty', in Richard T. Hull, ed., *Ethical Issues in the New Reproductive Technologies* (Belmont, CA: Wadsworth, 1990), 9.

[4]Lori B. Andrews, 'Alternative Modes of Reproduction', in Sherrill

Cohen and Nadine Taub, eds, *Reproductive Laws for the 1990s* (Clifton, NJ: Humana Press, 1989), 374-7.

[5]Gena Corea and Susan Ince report that in the United States in 1985, 42 out of the 54 clinics accepted only married couples ('Report of a Survey of IVF Clinics in the U.S.', in Patricia Spallone and Deborah Lynn Steinberg, eds, *Made to Order: The Myth of Reproductive and Genetic Progress* [Oxford: Pergamon Press, 1987], 140).

[6]Judith Lorber, 'In Vitro Fertilization and Gender Politics', in Elaine Hoffmann Baruch, Amadeo F. D'Adamo, Jr, and Joni Seager, eds, *Embryos, Ethics, and Women's Rights* (New York: Haworth Press, 1988), 118-19.

[7]There are comparable barriers to access to donor insemination. See, for example, Deborah Lynn Steinberg, 'Selective Breeding and Social Engineering: Discriminatory Policies of Access to Artificial Insemination by Donor in Great Britain', in Spallone and Steinberg, eds, *Made to Order*, 184-9.

[8]See Rebecca Shuster, 'Sexuality as a Continuum: The Bisexual Identity', in *Lesbian Psychologies: Explorations and Challenges*, edited by The Boston Lesbian Psychologies Collective (Urbana: University of Illinois Press, 1987), 56-71.

[9]See Thomas A. Shannon, 'In Vitro Fertilization: Ethical Issues', in Baruch et al., eds, *Embryos, Ethics, and Women's Rights*, 156-7.

[10]See Congregation for the Doctrine of the Faith, 'Instruction on Respect for Human Life in its Origin and on the Dignity of Procreation: Replies to Certain Questions of the Day' (Vatican City, 1987). This document, of course, also expresses many concerns about the treatment and destruction of embryos, arguments that are not evaluated here.

[11]Donald DeMarco, *In My Mother's Womb: The Catholic Church's Defense of Natural Life* (Manassas, VA: Trinity Communications, 1987), 156-7.

[12]Ibid., 157; cf. Ronald D. Lawler, 'Moral Reflections on the New Technologies: A Catholic Analysis', in *Embryos, Ethics, and Women's Rights*, 167-77.

[13]Janice G. Raymond, 'Fetalists and Feminists: They Are Not the Same', in Spallone and Steinberg, eds, *Made to Order*, 63.

[14]DeMarco, *In My Mother's Womb*, 147, emphasis added.

[15]Ibid., 159.

[16]Ann Pappert, 'In Vitro in Trouble, Critics Warn', *The Globe and Mail* (6 Feb. 1988), A1. For a comparable discussion of IVF success rates in France, see Françoise Laborie, 'Looking for Mothers You Only Find Fetuses', in Spallone and Steinberg, eds, *Made to Order*, 49-50.

[17]Pappert, 'In Vitro', A14.

[18]Linda S. Williams, 'No Relief Until the End: The Physical and Emotional Costs of In Vitro Fertilization', in Christine Overall, ed., *The Future of Human Reproduction* (Toronto: Women's Press, 1989), 120-38.

[19]'What You Should Know About In Vitro Fertilizaton', in *Our Bodies...Our Babies? Women Look at the New Reproductive Technologies* (Ottawa: Canadian Research Institute for the Advancement of Women, 1989); 'Current Developments and Issues: A Summary', *Reproductive and Genetic Engineering* 2, 3 (1989): 253.

[20]Anita Direcks, 'Has the Lesson Been Learned?: The DES Story and IVF', in Spallone and Steinberg, eds, *Made to Order*, 163. For a discussion of the harmful effects of one hormone used in IVF, clomiphene citrate, see Renate Klein and Robyn Rowland, 'Women as Test-Sites for Fertility Drugs: Clomiphene Citrate and Hormonal Cocktails', *Reproductive and Genetic Engineering: Journal of International Feminist Analysis* 1, 3 (1988): 251-73.

[21]See Harriet Simand, '1938-1988: Fifty Years of DES—Fifty Years Too Many', in Christine Overall, ed., *The Future of Human Reproduction* (Toronto: Women's Press, 1989) 95-106.

[22]Carl Wood and Ann Westmore, *Test-Tube Conception* (Englewood Cliffs, NJ: Prentice-Hall, 1984), 105.

[23]Klein and Rowland, 'Women as Test Sites', 270.

[24]'Resolution from the FINRRAGE Conference, July 3-8, 1985, Vallinge, Sweden', in Spallone and Steinberg, eds, *Made to Order*, 211.

[25]Sultana Kamal, 'Seizure of Reproductive Rights? A Discussion on Population Control in the Third World and the Emergence of the New Reproductive Technologies in the West', in *Made to Order*, 153.

[26]See, for example, Susan Sherwin, 'Feminist Ethics and In Vitro Fertilization', in Marsha Hanen and Kai Nielsen, eds, *Science, Morality and Feminist Theory* (Calgary: University of Calgary Press, 1987), 265-84.

[27]Margarete Sandelowski, 'Failures of Volition: An Historical Perspective on Female Agency and the Cause of Infertility', *Signs: Journal of Women in Culture and Society* 15, 3 (Spring 1990): 498.

[28]Ibid.

[29]Christine St. Peters, 'Feminist Discourse, Infertility and the New Reproductive Technologies', *National Women's Studies Association Journal* 1, 3 (Spring 1989): 359.

[30]Christine Crowe, '"Women Want It": In Vitro Fertilization and Women's Motivations for Participation', in *Made to Order*, 84-93.

[31]See Alison Solomon, 'Integrating Infertility Crisis Counseling into

Feminist Practice', *Reproductive and Genetic Engineering* 1, 1 (1988): 41-9; and Naomi Pfeffer, 'Artificial Insemination, In-Vitro Fertilization and the Stigma of Infertility', in Michelle Stanworth, ed., *Reproductive Technologies: Gender, Motherhood and Medicine* (Minneapolis: University of Minnesota Press, 1987), 81-97.

[32]Judith Lorber, 'Choice, Gift, or Patriarchal Bargain?', *Hypatia* 4 (Fall 1989): 30.

[33]Mary Anne Warren, 'IVF and Women's Interests: An Analysis of Feminist Concerns', *Bioethics* 2, 1 (1988): 40-1.

[34]Raymond, 'Reproductive Technologies', 133-42.

[35]Ibid., 135.

[36]Ibid., 137.

[37]Deborah Poff, 'Reproductive Technology and Social Policy in Canada', in Overall, ed., *The Future of Human Reproduction*, 223.

[38]Lorber, 'Choice, Gift, or Patriarchal Bargain?', 24.

[39]Vicki Van Wagner and Bob Lee, 'Principles into Practice: An Activist Vision of Feminist Reproductive Health Care', in Overall, ed., *The Future of Human Reproduction*, 238-58.

[40]Nikki Colodny, 'The Politics of Birth Control in a Reproductive Rights Context', in ibid., 43.

[41]'Current Developments and Issues: A Summary', *Reproductive and Genetic Engineering* 2, 3 (1989): 253, Wagner's emphasis.

[42]Lorber, 'Choice, Gift, or Patriarchal Bargain?', 23-6.

[43]Annette Burfoot, 'Exploitation Redefined: An Interview with an IVF Practitioner', *Resources for Feminist Research/Documentation sur la recherche féministe* 18, 2 (June 1989): 27.

Chapter 9

♦ ♦ ♦

Reproductive Engineering
and Genealogy

I shall begin this discussion of the genealogical implications of reproductive engineering and new reproductive technologies by describing in some detail the origins—the genealogy—of an infant to whom I shall refer as 'Baby X'.

Baby X was born in July 1987, to a woman named Heather Allen. Ms Allen had become pregnant in late 1986 after the transfer to her uterus of two embryos generated through *in vitro* fertilization undertaken at the Deer Falls Clinic in the United States. The sperm used was produced by James Travis, a man unrelated and unknown to Ms Allen. Mr Travis had sold his sperm to a commercial sperm bank, which had then provided the sperm to the IVF clinic where Ms Allen was impregnated. The eggs used for IVF in this case had been obtained from a woman called Cynthia Connor, also unknown and unrelated to Heather Allen. Ms Connor had been a participant in the IVF program at the Deer Falls Clinic one year earlier. After hormonal hyperstimulation, Ms Connor's ovaries had produced twelve eggs. Some were used for her own IVF attempts, but she agreed that several of them could be frozen for possible future use by other women. Four such eggs were used for Heather Allen's IVF attempt.

So Cynthia Connor's eggs, thawed after successful cryopreservation (freezing), were fertilized with James Travis's sperm, which was obtained through a national sperm bank. The two resulting healthy embryos were then transferred to Ms Allen's uterus; one successfully implanted, resulting in the birth, nine months later, of Baby X. However, after Baby X was born, she was removed through a standard adoption process to the custody of a man named Craig Banks and his common-law wife, Veronica Manchester. Then, in mid-1989, Mr Banks died. When Veronica Manchester subsequently married a man named Jason Abray, Mr Abray adopted Baby X.

As a result of these various reproductive technologies and practices, Baby X has six distinct parents, using the term 'parent' in several different senses: Cynthia Connor and James Travis are her genetic mother and genetic father, who provided the gametes from which she was generated; Heather Allen is her gestational mother, who gestated and gave birth to her; and Craig Banks, Veronica Manchester, and Jason Abray, who have contributed to rearing her, are her social parents.

Baby X is, of course, not a real infant. I have invented her life story in order to represent many—though by no means all—of the present variations in procreation and parenting made possible through 'high-tech' reproductive interventions. The case of Baby X provides a glimpse of some of the challenges that reproductive engineering is creating for future genealogical research.

Current reproductive technologies and practices include donor insemination (DI), the use of so-called 'fertility drugs', *in vitro* fertilization, the cryopreservation of gametes and embryos, prenatal diagnosis, sex selection and preselection, and so-called 'surrogate' or contract motherhood. This chapter presents some speculations about the implications of reproductive engineering and new reproductive technologies (NRTs) for future genealogical research. What challenges and problems do high-tech reproductive interventions generate for the genealogists and historians of the near and not-so-near future?

In attempting to answer this question, I shall focus not on outlining statistics and studies about reproduction, but rather on painting a picture of the sorts of things that are happening in reproductive science and the ways in which they might be important to family historians. While the emphasis here is on present-day practices, I shall conclude with a look at possible future developments. Since NRTs constitute an already huge and growing field, it is not possible to cover every new reproductive intervention; I shall merely indicate some of the main possibilities arising from changes in reproductive technologies and related reproductive practices. And while I shall take note of some morally problematic aspects of these practices, the aim of this chapter is not to evaluate them but to indicate their significance for future historical research.

♦ ♦ ♦
Current Social Conditions

It is important, first, for genealogists to be aware of the evolving social and medical conditions that are changing the context of reproduction within the Western world—indeed, that are helping to determine who reproduces, how they reproduce, when they reproduce, and how often they reproduce. These conditions include the apparently growing incidence of infertility; the development, theoretically at least, of more opportunities for single individuals, male and female, and for gay and lesbian couples, to become parents; the use of social screening to determine who does and who does not have access to high-tech infertility treatments; the use of prenatal diagnosis and sex preselection, along with abortion, to determine which foetuses will or will not be born; the use of fertility drugs that augment ovulation and thereby increase the likelihood of multiple foetuses; and the use of foetal surgery and aggressive measures to sustain the lives of foetuses and very premature infants.

The first of these new conditions affecting reproduction is the apparently growing incidence of infertility. It is estimated that as many as one in six couples suffers from some form of infertility, in the woman, in the man, or in both. While there is controversy as to whether the rate of infertility really is growing, or whether it is merely being recognized and reported with greater frequency than in former times, there can be no doubt that some factors contributing to the incidence of infertility are relatively new in human history. These factors include the increasing prevalence and variety of sexually-transmitted diseases; reproductive hazards in the environment generally and specifically in the workplace; and the use of prior reproductive interventions such as contraceptive hormones and appliances, including the intrauterine device (IUD).

Moreover, the condition of infertility is no longer simply accepted by women and men as an 'act of God' or a matter of 'fate'. Growing medical attention to infertility, along with the socially-created demands of infertile 'consumers', ensures that the scientific and medical establishments are devoting more and more attention to infertility. Unfortunately, however, most of this attention is directed not towards preventing infertility—a less painful and economically less costly approach—but rather towards attempting to 'fix' or 'cure' infertility, often by dramatic medical interventions, after it has occurred.

A second condition affecting reproductive practices is that these new reproductive technologies increase the opportunities, in theory at least, for single individuals to achieve biological and social parenthood. Of course, single parents are not new in human history: the death of a spouse, divorce, and more recently the deliberate choice by some pregnant women not to marry have made single parenthood more and more prevalent. Now, however, in theory at least, a woman alone can become a mother through the use of medically-supervised donor insemination, or, more informally, through self-insemination using sperm obtained from an acquaintance or a 'friend of a friend', or even through IVF using donated or purchased sperm. Likewise, in theory at least, it is possible for a single man to become a parent, by hiring a contract mother to be inseminated with his sperm. Similarly, there is now a greater possibility of parenting by same-sex couples: for example, a pair of gay men might hire a contract mother, have her inseminated with sperm from one of them, and raise the resulting infant. In a lesbian couple, one woman could be inseminated with donated sperm, and both women could raise the resulting infant.

I describe all of these scenarios as 'possibilities' rather than as prevalent conditions because of course certain social circumstances reduce the likelihood that unmarried individuals or gay or lesbian couples will actually be able to become parents through the use of these reproductive technologies. In the case of DI and IVF, physicians and clinics often reject, as a matter of straightforward discrimination, prospective single parents, and lesbian and gay parents. And in the case of both IVF and contract motherhood, the very high costs ordinarily involved make it less likely that women, who usually earn less than men whether they are single or in a same-sex couple, will be able to afford these procedures.

Thus a third condition affecting reproductive practice is the fact that, through a process of social screening and gate-keeping, current medical responses to infertility help to determine who reproduces and how. For example, access to IVF and donor insemination is usually limited by medical practitioners to those who are heterosexual and married or in a stable partnership, and who evince, in the eyes of medical practitioners, the allegedly appropriate moral characteristics for parenthood. (In some jurisdictions there is an ongoing debate about the ethical and legal legitimacy of making these distinctions.) Moreover, while IVF is sometimes covered by state-sponsored or private health plans, in most places it is extremely expensive for the individual consumer; in the United States, for example, IVF can cost up to $8,000 per attempt, and many attempts may occur before either a pregnancy is achieved, or, more frequently, the woman gives up. Hence few people outside the upper middle and upper classes can afford to pay for access to it.

Similarly, hiring a 'surrogate' or contract mother also requires extraordinary amounts of money—generally $25,000 to $30,000 or more. Since contract mothers themselves are usually relatively poor, uneducated, under-employed, or unemployed, in such arrangements those who are wealthy purchase the gestational services of working-class women. In general, then, social structures, not medical conditions, usually determine who does and who does not have access to the new reproductive practices and technologies.

A fourth condition affecting reproduction is that the use of various forms of prenatal diagnosis, such as ultrasound imaging, amniocentesis, and chorionic villi sampling (a test, performed in early pregnancy, of tissue from the placenta), now permits somewhat more successful reproduction by individuals who are carriers of genetically transmitted diseases. Prenatal screening can identify foetuses with such diseases as Down's syndrome, spina bifida, and Tay-Sachs disease; such foetuses may then be aborted, and the woman can become pregnant once again in the hope that the resulting foetus will be free of disabilities.

Another possible use (or misuse) of prenatal diagnosis is for purposes of sex selection. After prenatal screening, which ordinarily reveals the sex of the foetus, foetuses of the 'wrong' or undesired sex can be aborted. Similarly, scientists have invented methods of sex preselection, to attempt to determine the sex of offspring even before pregnancy has occurred. One such method involves the filtering of sperm specimens in order to separate out either the Y-bearing or X-bearing sperm, depending on whether a boy or a girl is wanted. The woman is then inseminated with sperm of the desired type; this procedure increases the likelihood of bearing a child of the desired sex, although it does not guarantee success. Still another method of sex preselection is to examine embryos produced through IVF to determine their sex, and then to implant in the woman's uterus only those of the desired sex.

Many sociologists believe that, given the preference for males in Western culture and indeed almost all cultures, the likely result of these practices, if widely adopted, would be the birth of more male offspring. (In India, for example, almost all of the foetuses aborted for reasons of sex selection are female.) Thus the practices of sex selection and preselection enable people to act upon their biases for and against children of one sex or the other, and possibly even to change the existing sex ratio in the population.

A fifth condition affecting procreation in North America is the growing use of so-called 'fertility drugs', chiefly hormones used to induce or augment ovulation so that more than one ovum or egg is produced. Fertility drugs may be used on their own, or in conjunction with *in vitro* fertilization. Fertility drugs used on their own increase the likelihood that a woman will become pregnant with more than one foetus. And although IVF has a very low success rate (about ten per cent, or fewer, of women who undergo IVF actually have babies as a result), successful IVF also results in a greater chance of multiple foetuses, because the usual practice is to implant several embryos, generated from the multiple ova, in the woman's uterus simultaneously.

Hence some forms of reproductive medicine result in the procreation of greater numbers of multiples—twins, triplets, quadruplets, and quintuplets. Occasionally a woman may become pregnant with six, seven, or even eight foetuses, although, needless to say, the infants in these grand multiple pregnancies do not survive; indeed, the mortality and morbidity rates for both mother and foetuses are also higher in the more usual cases of multiple gestation. Moreover, one form of reproductive intervention begets others: because of the dangers in grand multiple pregnancy for both the mother and the foetuses, a new procedure called 'selective termination' has been developed, in which some of the foetuses in the woman's uterus are killed in order to permit those remaining to survive and to reduce the health risks to the woman.

A final condition affecting reproduction is the availability of various high-tech interventions to ensure that some only marginally viable foetuses and infants survive pregnancy and the first few weeks of life. Aggressive measures are now being adopted to sustain the life of infants born after only twenty-three or twenty-four weeks of gestation, and 'foetal surgery' is sometimes undertaken during pregnancy to attempt to correct life-threatening conditions in the foetus.

All of these general conditions help to influence who reproduces, when, how, and with whom, and how many surviving offspring they will have.

♦ ♦ ♦
Some General Problems

Recent social practices have been challenging both our accepted ideas of relatedness and the traditional notion of the family as a unit composed of a husband, a wife, and their biologically related children. The use of new reproductive technologies extends those challenges. Is a mother the woman who gestates a child? The woman who provides the ovum from which the child was conceived? Or the woman who raises the child? Is the father the man who provides the sperm from which the child was generated? Or the man who raises the child?

Although in most cases today the social parents—the persons who raise the child—are still the same as the biological parents, new reproductive technologies and practices make possible the separation of genetic and social parenting, through the use of donated or purchased gametes (ova and sperm) and donated or purchased embryos. They also permit the separation of genetic and gestational mothering, so that the woman who gestates and gives birth to an infant need not be the woman from whose ovum the infant grew. These technological possibilities implicitly raise questions about the social and moral meaning of the genetic link between parents and offspring; they indicate that other forms of parenting are worthy of historical note; and they make research about family generation and formation more complicated.

For present and future historical researchers, new reproductive technologies and the practices associated with them are generating a number of pragmatic difficulties with respect to record-keeping. These difficulties generally fall into four categories: the non-existence or incompleteness of records, their accuracy, access to them, and their complexity.

First are the problems that may arise with respect to the very *existence* of records of some children's parentage. For example, in donor insemination it was formerly common to mix several sperm samples, obtained from different men, prior to the insemination of the woman, so that if conception took place it was difficult to know which man was the biological father of the resulting child. Although this practice is now being abandoned, the emphasis on anonymity associated with donor insemination meant that physicians in the recent past often kept very limited records of who their donors were, in order to guarantee that no one—including, or especially, the possible offspring themselves—would be able to ascertain the identity of the biological father.

Second, recent reproductive practices are generating some problems with respect to the *accuracy* of records. For example, the usual custom in most jurisdictions is for any husband who consents to his wife's insemination with donor sperm to be defined as the father of the resulting infant and listed in its birth record as its biological father, even though he has no genetic connection with the child. (Interestingly, efforts are made to obviate this practice in the case of contract motherhood, where the husband, if there is one, of the contract mother specifically disavows paternity of the resulting infant.) Similarly, informal anecdotal evidence suggests that in some cases of contract mother-hood in the United States, the gestational mother may register at the hospital where she gives birth using the name of the wife of the man who provided the sperm for the birth mother's insemination. The subsequent birth registration may then also falsely record the prospective social mother as the biological mother of the infant, even though she has no genetic connection to it.

A third problem arises in connection with *access* to the records of doctors, clinics, and hospitals: records that would indicate the true biological connections among parents and off-spring. Certainly records of donor insemination—especially the identities of donors—have usually been kept confidential, largely to protect the alleged interests of the donors themselves rather than those of the offspring or biological mothers.

In addition, it is reasonable to anticipate that the records of cryopreservation of embryos at IVF clinics will likewise be in-accessible to researchers. The result could well be difficulties for future historians in finding out whether an infant conceived through IVF is the offspring of a donor egg, for example, or whether two infants born at different times may in fact have been conceived at the same time, with one embryo being im-planted immediately and the other being frozen for later thawing and implantation.

A fourth problem for record-keeping is generated by the sheer *complexity* of the reproductive relationships made possible by new technologies. Among the specific complications arising from three of the main reproductive technologies and practices—donor insemination, *in vitro* fertilization, and cryopreservation—are the following.

Donor insemination ◆ Donor insemination can be used for women without a male partner, or whose male partner is infertile or has or carries a genetically transmitted disease. In addition to the problems arising from the mixing of sperm from several donors, the lack of adequate records, the preservation of donors' anonymity, and the inaccuracies on the birth certificate of DI off-spring, it should also be mentioned that it has been the custom, until very recently, for many donors to sell their sperm many times. This practice is facilitated through the cryopreservation of sperm and the creation of sperm banks with hundreds of sperm samples in the deep freeze. One result is that offspring born to different women within a city or a region may unknowingly all be half-siblings with the same biological father. (In one case of which I was told, a physician in Washington, DC, sold his sperm hundreds of times while he was at medical school. He is now married, and has advised his adolescent children not to marry anyone in Washington, DC, to avoid the possibility of marrying a half-sibling!)

There may also be cases in which a family member—for example, the biological brother or father of the social father of the infant—provides the sperm. In such a case the social father would be, strictly speaking, the biological uncle or brother of his child. In general it is very difficult for the DI child to discover the identity of his or her biological father. The practical problems are further compounded by the fact that the biological mother and her husband are often advised—wrongly, in my opinion—by physicians and psychologists not to disclose to their DI offspring the true story of their origins.

In vitro fertilization • Provided that accurate records are maintained and available (and this cannot be counted on), in its simplest form the procedure of *in vitro* fertilization is not likely to cause research problems for the genealogist. Nor are related procedures, such as gamete intrafallopian transfer (GIFT), inherently problematic.

However, complications could arise in some circumstances. For example, the couple may undergo IVF making use of donated or purchased eggs or sperm or even both, so that the resulting infant is related to only one or to neither of its subsequent social parents. The donated or purchased gametes may be obtained either from strangers or from within the biological family of the social parents. Thus if intrafamilial sperm donation can complicate relationships with respect to the men in the family, IVF also makes possible such complications with respect to the women.

Several noteworthy examples are on record, including a woman in her late forties in South Africa who gestated and gave birth to her own grandchildren. Embryos produced from the ova of her biological daughter and the sperm of her son-in-law were implanted in her uterus, and she bore triplets. She was thus the gestational mother and the genetic grandmother of the three children. In another case, a daughter bore a child 'for' her mother: she gestated an embryo produced from her mother's ovum and her mother's second husband's sperm. One can also imagine a comparable scenario in which a woman bears a child 'for' her biological sister, using an embryo produced through the *in vitro* fertilization of her sister's egg. Embryos for donation or sale to relatives or strangers can also be obtained through the controversial procedure called uterine lavage.

Cryopreservation • The practice of cryopreservation or freezing of extra or so-called 'spare' embryos generated through IVF also has the potential for complicating family relationships, since it produces an often large gap between the date of fertilization and the dates of implantation and birth. Perhaps a genealogist need not be concerned with the date of fertilization, in itself—certainly in 'ordinary' reproduction, the conception date is not normally known to researchers with any precision. However, cryopreservation has recently yielded at least one instance of what were called 'frozen twins', and this use may be of interest to genealogists. These female siblings were, technically, fraternal twins, but because of cryopreservation they were born at different times. One girl was the result of the immediate implantation of an embryo in her mother's uterus. The other girl, born several years later, was the product of an embryo produced at the same time but frozen and later thawed and implanted in her mother's uterus.

Cryopreservation has also generated so-called 'orphan embryos': that is, embryos whose biological parents are either no longer living or no longer interested in preserving or implanting them. In an early case in Australia, two embryos, produced from the eggs of a married woman and donated sperm, were 'orphaned' after the woman and her husband were killed in a plane crash. The fate of the embryos was considered of special significance since the couple was otherwise childless and left a very large estate; as a result, many women apparently offered to serve as gestators of the embryos.

In the more recent case of Mary Sue Davis and her husband Junior Davis (see Chapter 5) control of the embryos was eventually awarded jointly to the woman and the man. In many jurisdictions, however, the practice of cryopreservation continues to raise questions about the rights and responsibilities of the gamete-providers, the medical personnel, and the storage facility itself with respect to the use and disposition of frozen embryos.

◆ ◆ ◆

Possible Future Developments

In considering possible future developments in reproductive practices and technologies that could affect the work of the historian, it is important to remember that reproductive engineering is a subject of ongoing debate in countries such as Canada, the United States, Great Britain, and Australia. Depending on their content, the social policies developed within each country may serve to help or to hinder genealogical work. Family historians would therefore be well-advised to keep a close watch on the development of legislation on reproduction, particularly legislation with respect to the possible legalization or criminalization of contract motherhood, the development of or limitations upon *in vitro* fertilization and other techniques of so-called 'assisted reproduction', the future of sperm banks and the possible development of egg and embryo banks, and the determination of who should have access both to the technologies themselves and to medical and legal records about their use.

In the more distant future, some scientists and medical technicians are predicting possible developments in reproductive technology that will involve even more major interventions into human reproduction. Among the 'brave new world' scenarios that have been proposed are the following: the direct transplantation of foetuses from one uterus to another; ectogenesis, or the gestation of foetuses outside the uterus, using machines that mimic the functions of women's bodies; so-called 'male pregnancy', that is, the hormonal and/or surgical alteration of the male body to permit the implantation of one or more embryos in the abdomen and their subsequent gestation and delivery via Caesarean section; the cloning of human beings, to permit the exact genetic replication of individuals; parthenogenesis, or the generation of a new (female) human being solely from an egg cell; egg fusion, which is the generation of a female human being from the joining of two egg cells; and the production of chimeras—hybrids of human beings and animals.

While most of these proposals are highly controversial, and all of them sound like the wildest science fiction, it is important to remember that it was just a short time ago that today's reproductive interventions, such as *in vitro* fertilization and the cryopreservation of embryos, existed only in the realm of the human imagination. The high-tech reproductive fictions of today can very easily become the reproductive realities of tomorrow.

Bibliography

George Annas. 'Crazy Making: Embryos and Gestational Mothers'. *Hastings Center Report* 21, 1 (January/February 1991): 35-8.

Sherrill Cohen and Nadine Taub, eds. *Reproductive Laws for the 1990s*. Clifton, NJ: Humana Press, 1989.

Combined Ethics Committee of the Canadian Fertility and Andrology Society and the Society of Obstetricians and Gynaecologists of Canada. 'Ethical Considerations of the New Reproductive Technologies'. Toronto: Ribosome Communications, 1990.

Jonathan Glover et al. *Ethics of New Reproductive Technologies: The Glover Report to the European Commission*. DeKalb: Northern Illinois University Press, 1989.

Richard T. Hull, ed. *Ethical Issues in the New Reproductive Technologies*. Belmont, CA: Wadsworth, 1990.

Darryl Macer, Roger A. Balk, Benjamin Freedman, and Marie-Claude Goulet. 'New Creations?' *Hastings Center Report* 21, 1 (January/February, 1991): 32-5.

Ruth Macklin. 'Artificial Means of Reproduction and Our Understanding of the Family'. *Hastings Center Report* 21, 1 (January/ February 1991): 5-11.

Ontario Law Reform Commission. *Report on Human Artificial Reproduction and Related Matters*. Volumes I and II. Toronto: Ministry of the Attorney General, 1985.

Barbara Katz Rothman. *Recreating Motherhood: Ideology and Technology in a Patriarchal Society*. New York: W.W. Norton, 1989.

Michelle Stanworth, ed. *Reproductive Technologies: Gender, Motherhood and Medicine.* Minneapolis: University of Minnesota Press, 1987.

Mary Warnock. *A Question of Life: The Warnock Report on Human Fertilisation and Embryology.* Oxford: Basil Blackwell, 1984.

Index